Conserving Migratory Pollinators and Nectar Corridors in Western North America

Arizona-Sonora Desert Museum Studies in Natural History

SERIES EDITORS

Richard C. Brusca
Thomas R. Van Devender
Mark A. Dimmitt

Conserving Migratory Pollinators and Nectar Corridors in Western North America

Edited by Gary Paul Nabhan

Technical Editing by Richard C. Brusca and Louella Holter

The University of Arizona Press and
The Arizona–Sonora Desert Museum | Tucson

The University of Arizona Press
www.uapress.arizona.edu

First paperback edition 2020

ISBN-13: 978-0-8165-2254-5 (cloth)
ISBN-13: 978-0-8165-4242-0 (paper)

Cover illustrations: lesser long-nosed bat, white-winged dove, rufous hummingbird, monarch butterfly; © Paul Mirocha

Library of Congress Cataloging-in-Publication Data
Conserving migratory pollinator s and nectar corridors in western North America / edited by Gary Paul Nabhan.
p. cm. — (Arizona-Sonora Desert Museum studies in natural history)
Includes bibliographical references and index.
ISBN 0-8165-2254-5 (cloth : alk. paper)
1. Conservation biology—West (U.S.) 2. Pollinators—West (U.S.)
3. Animal migration—West (U.S.) 4. Corridors (Ecology)—West (U.S.)
I. Nabhan, Gary Paul. II. Series. QH76.5.W34C64 2004
333.95'16'0978—dc22
2003025331

Printed in the United States of America
⊗ This paper meets the requirements of ANSI/NISO Z39.48-1992 (Permanence of Paper).

Dedicated to William A. Calder, Dr. Hummingbird

Contents

Introduction

GARY PAUL NABHAN

This is a book of comparative zoogeography and conservation biology, one that considers the similarities and differences among the behavior and habitat requirements of several species of migratory pollinators and seed dispersers. Although other species are addressed in passing, the four species given greatest attention in this work are rufous hummingbirds, white-winged doves, lesser long-nosed bats, and monarch butterflies. We are particularly concerned with the population dynamics of these four species in the flyways that extend from the Pacific Ocean to the continental backbone of the Sierra Madre Occidental and Rocky Mountains. Most of the field studies reported here provide the greatest insights into the foraging and roosting behaviors of these four species as they move northward, from the Tropic of Cancer in western Mexico, into the deserts, semiarid grasslands, and subtropical thornscrub near the U.S.–Mexico border. Significantly, we are attempting to understand migratory populations of these species in both space and time; that is, we wish to know whether their densities, survival rates, and health are changing in response to alterations in the distribution and abundance of nectar plants found within their ranges.

Of the many conservation issues that have emerged over the past two decades, perhaps one more than any other signals a paradigm shift in the way applied ecologists now conceptualize and go about their work: the issue of how to safeguard the dynamic interactions among plants and pollinators. This issue is emblematic of a dramatic shift in conservation biology because it is no longer focused on protecting individual "things" (such as single species or single habitats); it is about maintaining interspecific relationships across a chain of habitats, so that landscape-level ecological processes continue to function (Nabhan and Fleming 1993).

Of course, the more general issue of conserving mutualisms or coevolved relationships between plants and animals is not a new concern among conservation biologists; it was implicit in classic papers by Paul Ehrlich, Peter Raven, Daniel Janzen, and Peter Feinsinger written in the 1970s and 1980s. Nevertheless, the protracted attention now given to a topic such as the conservation of plant-pollinator relationships by resource managers, policy-makers, environmental educators, and the press is, in fact, unprecedented. It began to surge with the onset of the multi-institutional "Forgotten Pollinators Campaign" launched in 1995, soon followed by the publication of a book, *The Forgotten Pollinators* (Buchmann and Nabhan 1996). Although the Arizona-Sonora Desert Museum was given much of the credit for this heightened interest in threatened interactions between plants and their pollinators (Kearns et al. 1998; Jacobson 1999; Withgott 1999), many other international organizations and institutions had already begun adding disrupted plant-pollinator interactions to their conservation agenda (Allen-Wardell et al. 1998; Kremen and Ricketts 2000). By the end of 2000, even the U.S. Departments of Agriculture and the Interior had collaborated on defining a shared research and management agenda for dealing with declining pollinators (Tepedino and Ginsberg 2000). As scientists began to "put a price on ecosystem services" (Daily 1997), the estimate that insect pollination of forty U.S. commercial crops is valued at $30 billion dollars per year was featured in a high-profile biodiversity report to U.S. decision-makers, prepared by the Smithsonian Institution and the President's Committee of Advisors on Science and Technology (Alonso et al. 2001).

The reasons for why this paradigm shift occurred when it did are complex, but farmers, florists, economists, and consumers realized that they had been receiving the "free" benefits of pollination provided by a diversity of vertebrates and invertebrates without investing in the maintenance of this ecosystem service (Abramovitz 1997; Nabhan and Buchmann 1997). When the U.S. Department of Agriculture declared that honeybee declines in the 1990s had triggered a "pollination crisis" of unprecedented proportions, a larger number of researchers whose work had been under-valued for decades suddenly had the attention of both journalists and policy-makers. They soon established that the question was not merely one of "how to save the honey-bee" but, more broadly, how best to maintain and restore the wildlands habitats that support the many pollinator species providing services essential to ecosystem health and food production (Nabhan 2001). To answer this question, detailed case studies of pollinators and their nectar resources would need to be undertaken at a landscape level over multiple seasons (Tepedino and Ginsberg 2000), or else the spatial and temporal variability in natural communities would mask the effects of intentional restoration. Fortunately, the Turner Endangered Species Fund and the Turner Foundation were willing to make a long-term commitment to funding research and management experiments over multiple years that will meet these criteria. Along with the generous support of other foundations—such as the C. S. Fund, the Wallace Research Foundation, the Wallace Genetics Foundation, the W. Alton Jones Foundation, the Kaplan Foundation, the Foundation for Deep Ecology, and the National Fish and Wildlife Foundation—our applied studies of pollination ecology were placed in the context of current theories advanced by conservation biologists regarding corridor design, habitat restoration, and population genetics. This book is the result of those collaborative efforts to bridge between field ecology and conservation biology, in theory and in practice.

At first glance, the contents of this book may look like a loose amalgam of case studies about very different animals that happen to visit the flowers of plants growing between the tropics of western Mexico and the deserts of the U.S. Southwest: three nectar-feeding bats, monarch butterflies, white-winged doves, and several hummingbirds. As one reads more deeply, however, it becomes clear that the researchers whose views are included in this book have been borrowing concepts, methodologies, and conservation perspectives from one another for some time, creating a synergy greater than the sum of its parts. In particular, many of the more recent studies have been directly or indirectly informed by Lincoln Brower's pioneering studies of butterflies and their co-evolution with plant secondary compounds, and by Theodore Fleming's detailed studies of bat-cactus mutualisms. Nevertheless, the Brower-Pyle chapter included here reminds us that certain geographic assumptions about monarch butterfly migration and their use of flowers during their long journeys have never been fully tested and are subject to question, given emerging evidence. Likewise, Fleming's chapter reminds us of the substantive gaps that remain in our understanding of nectar-feeding bat migration, despite all of the advances he and his colleagues have contributed over the past two decades. The late Bill Calder's review of rufous hummingbird migration, as well as the multi-authored chapter on white-winged doves, echo this theme: despite tremendous gains over the past decade in understanding these interactions, researchers remain humbled by how little they know relative to what they sense is *possible* to know or is *needed* in order to better conserve these relationships.

Yet we can marvel at how much new ground is finally being covered. The University of Guadalajara's team in the Sierra de Manantlan is at last informing North American ecologists about what hummingbirds and bats are doing during their winter stays in western Mexico. The Arizona-Sonora Desert Museum's team is documenting the phenological patterns of tubular flowers in sufficient detail to test hypotheses regarding the extent to which they shape the timing of hummingbird migration. My own work challenges the old assumption that Neotropical migrants are being most adversely affected in their tropical wintering grounds or in their northern breeding grounds; indeed, a more viable hypothesis may be that they are most stressed when passing through fragmented landscapes that aggravate the effects of drought on nectar availability. And Rodrigo Medellín's team challenges the self-serving assumption of some North American ecologists that stopover roosts for migrants in rural Mexico will not be valued and protected by local residents and therefore need "the scientific elite" to be in control of them.

All of the following chapters clearly demonstrate that a nectar trail is a migratory corridor distinct from those used by terrestrial carnivores or freshwater fish; it is a mosaic of stepping stones placed in a dissimilar matrix. Each species of migratory pollinator requires a different relationship between its stopover roost sites and nectar availability near those roosts; each migrant can tolerate conditions in the intervening and often hostile matrix between stopovers to varying degrees. The four migratory pollinators whose journeys form the essence of this story dramatically differ from one another in their foraging strategies and stopover site fidelities. Yet they all challenge many of the truisms that have emerged regarding the plight of migratory species in general. They can all be found in secondary as well as primary vegetation, and in some cases they continue to visit very urbanized, anthropogenic habitats. They are not necessarily put most at risk by the conditions they face in the tropics. And they all demonstrate some resilience in shifting from one stopover roost site to another when conditions dictate.

Nevertheless, they remain at risk for the many reasons elucidated in these chapters. Fortunately, as Medellín and his colleagues demonstrate, the more lasting collaborations among Mexican, Canadian, and U.S. environmental educators and conservation activists have already made a difference on the ground. Through bilingual, culturally adapted environmental education programs, tens of thousands of schoolchildren are now learning about these energetic pollinators and the flowering plants they depend on and are urging local land managers to take better care of them. Farmers and ranchers are being engaged in habitat enhancement and restoration projects to give these animals more safe refuges. And the U.S. and Mexican governments have recently decreed along the West Coast corridor several new and ecologically significant biosphere reserves, national monuments, and wildlife refuges, thereby diversifying the "safe sites" available to pollinators in this mega-region.

We hope that this book will both inform and inspire several more tangible conservation efforts along this line, such as the ones promoted by the Corredor Colibri, the Sky Islands Alliance, the Sonoran Institute, and the Center for Sustainable Environments (Nabhan 2001; Vanderpool 2002). These organizations have begun to restore pollinator corridors on farms and ranches in the U.S.–Mexico borderlands. Ultimately, we should measure the success of this and other collaborations not merely by how many more scientific articles they generate but by whether they truly secure a brighter future for those creatures that travel the nectar corridors between the Sierra Madre Occidental of Mexico and the Intermountain West of the United States and Canada. We will have failed if our concerns only generate more detailed obituaries and autopsies for devastated places, plants, and animals instead of leading to the restoration of healthy relationships among migrant pollinators, sessile plants, and the habitats that support them.

LITERATURE CITED

Abramovitz, J. N. 1997. Valuing nature's services. Pp. 95–114 in L. Stark, ed., State of the World 1997. W. W. Norton, New York.

Allen-Wardell, G., P. Bernhardt, R. Bitner, A. Burquez, S. Buchmann, J. Cane, V. Dalton, P. Feinsinger, M. Ingram, D. Inouye, C. Jones, K. Kennedy, P. Kevan, H. Koopowitz, R. Medellín, S. Medellín-Morales, G. P. Nabhan, B. Pavlik, V. Tepedino, P. Torchio, and S. Walker. 1998. The potential consequences of pollinator declines on the conservation of biodiversity and stability of food crop yields. Conservation Biology 12:8–17.

Alonso, A., F. Dallmeier, and P. Raven. 2001. Biodiversity: Connecting with the Tapestry of Life. Smithsonian Institution and President's Committee of Advisors on Science and Technology, Washington, D.C.

Buchmann, S., and G. P. Nabhan. 1996. The Forgotten Pollinators. Island Press, Washington, D.C.

Daily, G. 1997. Nature's Services. Island Press, Washington, D.C.

Jacobson, S. K. 1999. Communications Skills for Conservation Professionals. Island Press, Washington, D.C.

Kearns, C. A., D. W. Inouye, and N. M. Waser. 1998. Endangered mutualisms: The conservation of plant-pollinator interactions. Annual Review of Ecology and Systematics 29:83–112.

Kremen, C., and T. Ricketts. 2000. Global perspectives on pollinator disruptions. Conservation Biology 14(5):1226–29.

Nabhan, G. P. 2001. Nectar trails of migratory pollinators. Conservation in Practice 2(1):20–27.

Nabhan, G. P., and S. Buchmann. 1997. Services provided by pollinators. Pp. 133–50 in G. Daily, ed., Nature's Services. Island Press, Washington, D.C.

Nabhan, G. P., and T. E. Fleming. 1993. Conservation of New World mutualisms. Conservation Biology 7:457–59.

Tepedino, V. J., and H. S. Ginsberg. 2000. Report of the U.S. Department of Agriculture and the U.S. Department of the Interior Joint Workshop on Declining Pollinators. Information and Technology Report USGS/BRD/ITR-2000-007. Patuxent Wildlife Research Center, Patuxent, Md.

Vanderpool, T. 2002. Borderline hope: The corredor colibri. Tucson Weekly 19(3):20–24.

Withgott, J. 1999. Pollination migrates to the top of the conservation agenda. BioScience 49:857–62.

Conserving Migratory Pollinators and Nectar Corridors in Western North America

CHAPTER 1

Stresses on Pollinators during Migration

Is Nectar Availability at Stopovers the Weak Link in Plant-Pollinator Conservation?

GARY PAUL NABHAN

If little is known about the contingencies facing migrants on their wintering grounds, even less is known about the challenges they face along their migratory lanes. —D. H. Morse (1980: 451)

The conservation of migratory animals has been of concern to conservationists for more than a century, but our understanding of the causes and consequences of changes in migrant populations continues to undergo considerable revision. In the 1980s, when many conservation biologists and activists were expressing deep concern that populations of migratory landbirds, bats, and butterflies were declining because of land-use changes in Latin America, it was assumed that deforestation was eliminating, degrading, or fragmenting these migrants' wintering habitats (Terborgh 1980; Pyle 1983; Heacox 1989). These warning cries generated a tremendous effort in field research and policy initiatives to positively affect the status of pollinators, insectivores, frugivores, and predators that migrate between Central and North America (Brower and Malcolm 1991; Nabhan and Fleming 1993; Stotz et al. 1996; Arita and Santos del Prado 1999). However, it soon became clear that not all Neotropical migrants of concern were actually threatened with extinction, nor were they necessarily declining due to anthropogenic vegetation change in their Latin American wintering grounds (Hutto 1982; Cockrum and Petryszyn 1991; Malcolm 1987).

For instance, continent-wide population trends for migratory landbirds between 1969 and 1991 indicated that only twenty-three of fifty-three species tracked in the National Audubon Society's Breeding Birds Survey suffered net declines for the entire survey period. Notably, only four of those migrants suffered consistent unidirectional declines, as opposed to population fluctuations (Peterjohn et al. 1995). Of the 422 landbirds assessed by Stotz et al. (1996), only fifty species suffered population declines severe enough for them to be considered at risk of extinction. When experts on migratory landbirds, bats, or butterflies are now called upon to interpret the causes of any specific population decline or fluctuation, they no longer adhere to the simple dichotomy proposed by Terborgh (1989) in his polemic, *Where Have All the Songbirds Gone?*—that declines are the result of either wintering-ground deforestation or summer breeding-ground fragmentation.

Instead, contemporary conservation biologists are entertaining a variety of hypotheses to account for the population changes documented for Neotropical migrants: cowbird predation on bird eggs, the non-target effects of toxic pollen and herbicides of genetically engineered corn on monarch larvae, the dynamiting of nectarivorous bat caves by ranchers worried about vampire bat predation on their cattle, and global climate change, to name a few (Stotz et al. 1996; Nabhan 1999). Whether generated by climatic variability, herbicides, pesticides, or land conversion, stresses on pollinators during their migration are now being scrutinized just as much as those generated in their summering and wintering grounds.

One reason for paying more attention to the status of these species while they are in transit is that pollinators require a tight synchrony between the timing of their migration and the peak nectar availability of flowering plants along the corridors they travel (Fleming 2000). This synchrony can easily be disrupted by climate change or by anthropogenic vegetation change, leaving pollinators high and dry (Inouye et al. 2000).

The following discussion pays particular attention to the vulnerabilities of four migratory species, given the accumulating evidence of local or regional declines of these pollinators and seed dispensers: western populations of rufous hummingbirds *(Selasphorus rufus)*; the Sonoran Desert populations of white-winged doves *(Zenaida asiatica mearnsii)*; coastal populations of lesser long-nosed bats *(Leptonycteris curasoae)*; and monarch butterflies *(Danaus plexippus)* that migrate to the Transvolcanic Belt of Mexico from west of the Rio Grande watershed (Buchmann and Nabhan 1996; Pyle 1999; Nabhan 2001). Although these four species are not necessarily endangered in all or a significant portion of their range, their susceptibility to land-use changes brings into question the future stability of their pollination services (Buchmann and Nabhan 1996). The authors of this volume are currently evaluating whether or not significant population declines are occurring for these four species of migratory pollinators and for other hummingbirds as well. Regardless of the severity of population changes found for these particular migratory pollinators, it is clear that we understand little about their status at stopover sites during migration or about the anthropogenic stresses they face while in transit. After defining the terms necessary for understanding the migration of pollinators, I review evidence that migrants may be particularly vulnerable during long migrations across arid lands, especially if their stopovers are negatively affected by land-use changes. I then highlight a recent pollinator conservation collaboration along one migratory corridor that appears to have applicability elsewhere.

Defining the Ecological Role of Pollinators Along a Corridor

The concept of migratory corridors typically conjures up images of continuous linear habitats or greenways that provide for the movements of large predators and other wide-ranging species. In contrast, migratory corridors for winged pollinators might be more aptly described as a mosaic of stepping-stones within a larger matrix, with each stone a stopover that migrants use for "refueling" while in transit along 2,000–6,000 km flyways. The "glue" providing the connectivity in this mosaic is the shared presence of certain flowering plant genera that these mobile pollinators consistently visit if in bloom.

For example, lesser long-nosed bats use dense stands of columnar cacti, agaves, and morning glory trees—usually, but not always, situated near cave roosts—as their stepping-stones on their northward flight from Jalisco to southern Arizona. Many of the nectar-producing plants visited by long-nosed bats are patchily distributed succulents that favor hot, rocky hillsides and cliffs. The distances between these patches may be a limiting factor for nectarivorous bats, just as we know the availability of roosts in caves and rock shelters may be. If this hypothesis is confirmed, it may indicate that migratory pollinators such as long-nosed bats have always had to move considerable distances to find suitable stopovers, even before the intervening matrix was degraded.

Bat ecologist Donna Howell (1974) may have been the first to implicitly suggest the concept of a nectar trail—that is, a sequence of flowering plants situated around each stepping-stone along a migration route: "It is not uncommon to find several bat-pollinated species in association [with one another at the same site] exhibiting similar phenologies. . . . It appears, superficially at least, that these species compete for the services of pollinating bats" (312). From winter through late spring, these clusters of bat-pollinated plants bloom sequentially from south to north, creating the effect of a blooming wave cresting northward (Fleming 2000). Near-simultaneous blooming of several nocturnally flowering species at the same site has the effect of presenting a concentrated energy source to nectar-feeding migrants, which keeps them at a particular stopover roost until the nectar resources there begin to decline. The pollinator population then moves northward to seek the next emerging bloom in the northward-reaching wave.

A "nectar trail" is now envisioned to be the entire circulation pattern that pollinators follow as they migrate from one sequentially blooming plant population to the next (Fleming 2000). The loosely co-evolved relationships between migratory pollinators and plant populations contributing to the blooming wave may be thought of as "sequential mutualisms." Should one or more of the plant mutualists be eliminated from the sequence by any factor—habitat destruction, aberrant weather, competition, pests, or diseases—the nutritional status and movements of the pollinator may be disrupted to the extent that the animal cannot effectively visit other mutualists.

The concept of "sequential mutualisms" implies, for migratory pollinators, that an animal may be linked in space and time with several flowering plant populations. In the case of lesser long-nosed bats, a migratory population temporarily located at one roost may move pollen and fruit seeds between populations within a 100 km radius of that roost. Because the plants are sessile but the pollinators are not, nectarivorous bats, hummingbirds, doves, butterflies, and moths serve as "mobile links" among plant populations in different landscapes, facilitating pollen and gene flow over considerable distances. Similarly, lesser long-nosed bats and white-winged doves also facilitate pollen and seed dispersal as well as spatial mixing of genotypes from geographically isolated populations. They too serve as mobile links between plant populations, in this case, during two different phases in the life cycle of columnar cacti.

Although migratory pollinators ensure landscape-level linkages among many different plant populations, many non-migratory pollinators (such as honeybees) visit these same flowers and benefit secondarily from genetic mixing stimulated by the migrants. Should the plant populations linked by pollinators fall within officially designated parks, biosphere reserves, wildlife sanctuaries, or other protected areas, these migrants have special conservation significance. They may be among the few "mobile links" of any kind that visit most or all units in a regional reserve network, and this fact distinguishes them from the carnivores that are often proposed as the umbrella species to be used in designing such networks (Soule and Terborgh 1999).

Assessing Where Stresses Are Most Severe

In assessing the stresses that may lead to declines in migratory pollinators, it is obvious that some are generated from a single point source (such as destruction of a roost site) whereas others have a more pervasive influence (global climate change, or the spread of invasive species competitive with nectar plants). The more pervasive stresses may affect pollinators with greater severity during one life stage (during gestation or long-distance migration) than during others, because energetic costs and reproductive risks may be more pronounced during that life stage. In short, an interaction exists between the relative vulnerability of a migrant during a particular life stage and the habitat quality or resource availability of the habitat it is occupying.

Migratory species vary somewhat in the life stage during which they are most vulnerable. Lesser long-nosed bats may be pregnant when they make their northward migration across the desert in the spring, whereas rufous hummingbirds and white-winged doves are not but still need to maintain both weight and speed to obtain adequate breeding territories and nesting areas in the spring. In each case, the additional energetic demands of long-distance migration place these species under further stress. Underscoring the relatively acute vulnerability of migratory birds, Moore and Simons (1992) concluded that "the single most important constraint during migration is to acquire enough food to meet energetic requirements, especially for long-distance migrants which must overcome geographic barriers" (348). When the geographic barrier is a desert of relatively low nectar productivity

TABLE 1.1. Energy Costs for Migratory Pollinators

	Rufous Hummingbirds	Lesser Long-Nosed Bats
Body length	10 cm	9 cm
Migration distance (straight line south-north)	4,900 km (Jalisco-Alaska)	1,200 km (Jalisco-Arizona)
Distance in body length equivalents	49 million	13.3 million
Probable duration of one-way migration	36 days	60–85 days
Total fat cost of migration from Jalisco	9 g (to Alaska)	13 g (to Arizona)
Per km cost per gram of fat at 52 km per hr flight speed	544 km	92 km
Flowers needed per one-way migration	30,660 fuchsia flowers	227 cactus flowers

(discussed further below), long-distance migration across it may be particularly stressful.

Avian ecologists have estimated that, prior to departure, long-distance migrants must deposit fat (lipid) reserves averaging 30–50 percent of their body mass, compared to only 13–25 percent for short-distance migrants (Moore and Simons 1992). Table 1.1 compares the energy costs for the migration of rufous hummingbirds (see chapter 4 in this volume) with those for lesser long-nosed bats (Fleming, personal communication). Both of these species depart seasonally from moist tropical deciduous forests and move through dry subtropical thornscrub and hyperarid deserts before arriving in temperate woodlands, grasslands, coniferous forests, or other more northerly habitats. As their fat reserves are rapidly depleted during long-distance movements, migrants are forced to forage one to several days at stopovers to replenish lipid reserves (Lavee and Safriel 1989). Migrants that must dwell longer at stopovers risk delayed arrival at summer breeding or birthing grounds (Safriel and Lavee 1988; Moore and Simons 1992). If they arrive after inter- or intra-specific competitors are well established in their shared summering grounds, they may be relegated to territories or roosts of marginal quality. Reduced food availability, greater susceptibility to predators, or adverse weather conditions may ultimately reduce their fitness, or that of their progeny.

Although most of the principles discussed above are derived from studies of migratory insectivores, recent studies confirm that migratory pollinators such as rufous hummingbirds are severely disadvantaged by declining quality at stopover habitats, particularly by reduced nectar availability (Russell et al. 1994; Calder 1997). Heinemann (1992) documented how rufous hummingbirds in New Mexico abandon floral patches when average daily nectar availability dips below 0.2–0.3 ml per flower, because the remaining nectar supply cannot meet their minimum energetic requirements for foraging under those conditions. Abandonment typically occurs 1–3 days after available nectar resources drop below this threshold, regardless of whether this decline is the result of competition from bees and wasps (Heinemann 1992) or lessened productivity as a result of drought (Russell et al. 1994).

In desertscrub vegetation in California, Russell et al. (1994) confirmed the effects of drought on stopover habitat quality. During drought years, they witnessed selection against the leanest rufous hummingbirds arriving at the stopover they monitored. There was differential mortality of birds weighing less than 3.0 g, and even those weighing less than 3.2 g might not have survived if they did not find greater floral resource availability at their next stopover. These pollination ecologists recorded positive correlations between summer rainfall, habitat quality, and rufous hummingbird numbers at stopover sites. The drought years of 1981 and 1984 led to declines, and the wet years associated with El Niño Southern Oscillation (ENSO) effects in 1982 and 1983 led to population recoveries.

Curiously, it appears that, in at least some years, rufous hummingbirds cross from Sonora into California by island-hopping between Tiburón, San Esteban, San Lorenzo Sur, and Ángel de la Guarda (Case and Cody 1983: appendix 8.8), rather than proceeding up the Sonoran coast through the hyperarid Gran Desierto, then crossing land over to California. Although these islands are also hyperarid in the interior, the coastal fog effect may support flowering shrubs even in drought years. Nevertheless, island-hopping requires that hummingbirds heading to Baja California fly 60–150 km across water with no nectar available at all.

Fluctuations of some migratory bird populations have been positively correlated with stochastic variation in annual rainfall (Blake et al. 1995), but it remains unclear whether among-year annual precipitation variability can be used as a direct and reliable predictor of nectar availability. Drought, for example, may affect several components of nectar production to varying degrees: species composition (including the number of animal-pollinated species relative to wind-pollinated or selfing species); the abundance of individual annual wildflowers, flowering shrubs, or succulents; the total number of flowers per plant per season; nectar production per flower; duration and intensity of the flowering season; and evaporation rates of nectar in open flowers. Competition between annuals and perennials may also influence floral abundance, and the floral productivity of perennial plants may be influenced to some extent by droughts during previous years. Despite these cautionary caveats, variation in annual precipitation is still used as a surrogate for variation in nectar availability as it affects habitat quality (Russell et al. 1994; Blake et al. 1995).

Contrary to popular belief, rainfall at sites in desert biomes is not necessarily more variable (less predictable) than rainfall in non-desert biomes (Davidowitz 1998). Nevertheless, Davidowitz determined from analysis of 328 weather stations in western North America that the Lower Colorado River Valley subregion of the Sonoran Desert and the adjacent Mojave Desert are significantly more variable in both among-year annual precipitation and warm-season precipitation than are the Great Basin and Chihuahuan deserts, the Arizona Uplands subregion of the Sonoran Desert, and adjacent semidesert grasslands and woodlands. Warm-season precipitation had a larger range of variability from site to site and year to year than cold-season precipitation in southwestern North America, especially in low-elevation reaches of the Sonoran Desert dominated by creosotebush (Davidowitz 1998). At the heart of this most unpredictable subregion is the Gran Desierto of Sonora, which averages less than 50 mm/yr of rainfall at its lowest elevations and suffers no measurable precipitation for as much as 26 months running. Ezcurra and Rodrigues (1986) have argued that this area's extreme unpredictability of rainfall is probably of greater significance to plant life than are its averaged annual precipitation estimates.

Assessing the Relative Availability of Nectar Resources

The dominant perennials characteristic of the Lower Colorado River Valley subregion's creosote flats are estimated to produce 1.04–2.3

kg/ha of nectar sugar in the most favorable years of precipitation, compared to 4.4 kg/ha produced the same years within more complex arborescent vegetation of bajada slopes, characteristic of the Arizona Uplands subregion (Orians et al. 1977). In other words, there may be more than a four-fold difference in nectar resource availability between these two vegetation types within the Sonoran Desert, even when both enjoy exceptionally wet years. Such differences may be aggravated during drought years by the lesser predictability of the rainfall in the Lower Colorado River Valley and Central Gulf Coast subregions of the Sonoran Desert as compared to the Arizona Uplands subregion and adjacent grasslands, woodlands, and thornscrub.

Although the effects of rainfall variability on nectar production per hectare are complex, the relative importance of rainfall variability to the timing of floral anthesis is more direct. Bowers and Dimmitt (1994) determined that the flowering of ocotillo *(Fouquieria splendens)*, a plant pollinated by hummingbirds and carpenter bees, is triggered by cool-season rains of 11–46 mm. A follow-up study by Carmondy (1999) showed a highly significant (0.021 level) correlation between the time of floral anthesis for ocotillo and rainfall in the preceding month. The timing of rainfall was responsible for 34 percent of the variation in flowering time among years, whereas diurnal temperature ranges in the preceding month also exerted considerable control on when buds had sufficient degree-days to flower. Although genetically controlled responses to degree-days and moisture availability may be the proximate triggers to ocotillo's flowering, Waser (1979) has argued that hummingbird pollinator availability may be the evolutionary force that serves as the ultimate determinant of flowering time for this species. Seed set of ocotillo varied considerably from year to year in his southern Arizona plots, and this variation was correlated with the temporal match between flowering period and hummingbird arrival times at stopovers.

Ocotillo is not the only hummingbird-pollinated flower whose anthesis is influenced significantly by the timing of rainfall. Flowering times for the desert-dwelling coyote tobacco *(Nicotiana obtusifolia)* and Parry's beardtongue *(Penstemon parryi)* are also significantly correlated with the timing and amount of rainfall in the month preceding anthesis (Carmondy 1999).

The timing of initiation and the intensity and duration of spring flowering for these and other desert species (*Justicia* spp., *Salvia* spp.) may dramatically affect the fitness of long-distance migrants such as rufous hummingbirds. Their mean body mass when captured heading northward in the spring is 3.27 g, compared to a mean of 3.44 g when heading south in the fall (data from Russell et al., in Calder 1993). This suggests more severe nectar scarcity in the spring—when they typically follow a more coastal corridor through the Central Gulf Coast of Sonora and Lower Colorado River Valley subregions—than in the fall, when they pass from coniferous forests into oak woodlands, grasslands, and, marginally, Arizona Uplands desertscrub (Calder 1993). The mean body masses for breeding males (3.40 g) and females (3.58 g) surpass the mean for both sexes during spring migration, suggesting that food resources are more available during the breeding season and/or that energy expenditures may be less (Calder 1993). Mean body mass for males wintering in Jalisco is as low as that of spring migrants (3.27 g), but females in Jalisco may be somewhat more robust (3.35 g).

Though preliminary, these data appear consistent with the hypothesis that spring migration may be the most energetically stressful time for migrants such as rufous hummingbirds. This may be particularly true as they pass through the most nectar-scarce and unpredictable link in their migratory chain: from the Central Gulf Coast of Sonora, through the Lower Colorado River valley, and northward into the Mojave Desert. As A. H. Miller (1963) reported from Joshua Tree National Monument, a 2.5 g rufous hummingbird had its ability to remain in motion "fail during a migratory flight in hot sun through an area without water or nectar sources" (672). He was unable to revive the bird, which was well below the previously mentioned threshold weight of 3.0 g ensuring short-term survival.

I have used rufous hummingbirds to demonstrate how vulnerable long-distance migrants are to stopovers where nectar resources are poor or unpredictable, but this same issue may affect other migratory pollinators to varying degrees. Fleming (2000) reported little variation in flowering time and intensity in bat- and dove-pollinated saguaros, but substantial variation in the peak period and intensity of cardon cactus in response to variable weather. Less is known about the relative degree of variation in nectar availability in the composites and milkweeds used by adult monarch butterflies, and in the agaves and open-flowered trees used by lesser long-nosed bats. If we are to better understand how these pollinators are affected by natural stochastic processes, we need more detailed longitudinal studies of the effect of precipitation on the flowering intensity of plants visited by bats, hummingbirds, doves, and butterflies. There are few good phenological studies (e.g., Bullock and Solis-Magallanes 1990) accomplished within the west Mexican coastal corridor that can be used to assess the relative nectar availability at various stopover sites farther south.

Anthropogenic Stresses at Desert Stopovers

Given that year-to-year variability in nectar availability at desert stopovers naturally induces some pollinator mortality, what anthropogenic stresses further reduce nectar availability? One of the most severe of these stresses on native desertscrub habitats is the intentional planting—and subsequent escape—of exotic buffelgrass *(Pennisetum ciliare)* (Búrquez-Montijo et al. 2002). Buffelgrass has already been planted on 2.4 million ha of land in Sonora, and at least another 5 million ha have been planted in adjacent states. As in Sonora, it has also invaded beyond intentional plantings adjacent Sinaloa, Arizona, and Baja California. Between Benjamin Hill and Guaymas, Sonora intentionally planted and inadvertently escaped buffelgrass now constitutes more than 25 percent of the vegetation along arroyo margins, replacing many native nectar plants that formerly grew there (Búrquez-Montijo et al. 2002). Dense buffelgrass stands along washes also carry wildfires that kill burn-intolerant trees and columnar cacti that otherwise serve as roosts and provide nectar for migrants.

Although it is less invasive in the driest portions of the Sonoran Desert, buffelgrass is already present from Bahía de Kino through the Pinacate and Organ Pipe, where its direct competitive effects on ocotillos, hummingbird bushes, and beardtongues demands further study. Tragically, buffelgrass is just one of 350 exotic plant species now naturalized in the Sonoran Desert, with dozens of these exotics now competing with native floral resources along the migratory corridors (Nabhan 2002).

Lesser long-nosed bats are vulnerable to other threats where they congregate in large numbers at their maternity roosts (Arita and Santos del Prado 1999). Because they are mistaken for vampires or the mythic *chupacabras* (goat-suckers), they are often killed en masse by rural residents who burn, dynamite, or otherwise disturb their roosts. Monarch butterflies are also vulnerable because of their tendency to aggregate (Buchmann and Nabhan 1996), and this tendency occurs not merely at the overwintering sites but also when they are in transit. Pyle (1999) described the "oasis effect" in which monarch butterflies migrating through the Great Basin Desert aggregate in lone trees, using them as stopovers even where nectar resources appear to be limiting. Perhaps migrating monarchs carefully select transitory roosts for their microclimatic conditions much as they do their overwintering roosts (Weiss et al. 1991). These trees may be targets for woodcutters in the many rural communities where farmers and ranchers rely on fuelwood for 80 percent of their cooking and heating needs.

The severity of direct harvesting of saguaro cacti for their wooden "ribs" (used in furniture and fence building) and of wild agaves (used for bootleg mescal) deserves further investigation near major stopovers. Elsewhere, I have estimated that, in Sonora alone, between 500,000 and 1,200,000 wild agave ramets are harvested each year prior to the flowering of their paniculate stalks (Nabhan and Fleming 1993). If intensively undertaken within a 20 km radius of major maternity roosts, such harvesting may place further energetic stress on pregnant female populations of lesser long-nosed bats. Although much of the mescal harvesting is really selective pruning of multi-rosette clones (ramets) rather than outright destruction of genetic individuals (genets), its probable net effect is to reduce the genetic variation of "wild" agave populations within human reach.

Undoubtedly, the most irrevocable anthropogenic pressures on stopover habitats are the outright clearing, conversion, degradation, and fragmentation of wildlands habitats in urban and agricultural areas within the Sonoran Desert and adjacent coastal thornscrub (Nabhan and Holdsworth 1999). Although all the migrants under consideration here do use secondary vegetation (Hutto 1982) and have been found in urban areas, we know little about the minimum patch size of nectar-providing vegetation that they need to survive under these conditions (Lavee and Safriel 1989). Nevertheless, ecological restoration of 5–50 ha "stopover" patches of native vegetation may reduce these negative impacts, allowing recolonization of anthropogenically disturbed habitats by pollinators, as the case study below suggests.

Particularly in arid and dry subtropical landscapes, farmlands found between protected areas can serve either as oasis-like stopovers for these migrants (Lavee and Safriel 1989) or as barren, chemical-ridden sites that further stress pollinators during the most energy-intensive phase of their annual cycle (Lavee and Safriel 1989; Pyle 1999). Over the past half century, millions of hectares of desert and thornscrub vegetation in western Mexico and the U.S. Southwest have been converted to field crops or pasture grasses intensively managed with agrochemical grasses, creating 100–200 km stretches of flyways of chemically fragmented habitat largely devoid of suitable forage and roost sites for nectarivores. We are only beginning to fathom the long-term effects on migratory bats, doves, hummingbirds, and butterflies of having fewer nectar plants for forage and fewer safe roost sites available as stopovers.

Migratory pollinators are not the only migrants affected by physical and chemical fragmentation of their flyways. More than 70 percent of all birds, bats, and butterflies that migrate between the United States and Mexico travel routes bounded by the continental divide in the Sierra Madre Occidental and Rockies to the east, and by the Colorado River and Sea of Cortés to the west (Nabhan and Donovan 2000). Because habitat loss has an impact on so many species of migrants, ecological restoration aimed at restoring stopover areas for migratory pollinators may also positively influence other migrants and other non-migratory pollinators. The following case study outlines one such effort.

Restoring Ecological Connectivity

The best way to ensure adequate connectivity in regional reserve networks may be to better manage intervening private lands in a manner consistent with the needs of migratory wildlife. Yet, in their current state, many private lands are the weak links in the migratory chain. Restoring the ecological connectivity of these lands will require stronger stewardship collaborations among public agencies, private land owners, and rural ejido collectives.

Dr. Exequiel Ezcurra (formerly the lead scientist for Mexico's Instituto Nacionál de Ecologia) echoed this point in a keynote address remembered for its political wisdom as well as its excellent science. In May 1998, at the International Conference on the Conservation of Migratory Pollinators and Their Corridors, held at the Arizona-Sonora Desert Museum, he pointed to the increasing political difficulties of establishing additional large, federally protected areas in Mexico and the United States. He predicted that few new government-funded reserves are likely to be established in northwestern Mexico, so that restoring ecological connectivity through private lands between federally protected areas will be critical to binational regional conservation efforts.

One success story of public-private collaboration is the remarkable recovery of riparian corridors using treated sewage effluent along binational riverbeds in the Arizona-Sonora borderlands (Nabhan 2001). Because of its southeast-northwest alignment contiguous to north-south running rivers in Sonora, the Rio Santa Cruz is part of a 400 km corridor of intermittent streams and associated riparian vegetation stretching across some of the driest portions of arid North America. This corridor and that of the San Pedro and San Simon Rivers have unparalleled importance to binational wildlife movements, given that only 10 percent of the historic riparian vegetation remains along the rivers and streams of southern Arizona (Nabhan and Donovan 2000).

In 1980, the Nogales International Waste Treatment Plant began to augment the historically diminishing instream flow with treated effluent. The plant now provides continuous flow and replenishment of the shallow aquifer below the floodplain for 40 km north of Nogales, Sonora. By 1992, along a stretch of floodplain that had formerly lost most of its gallery forests, newly established stands of cottonwoods, willows, and mesquites covered more than 45 percent of the upper Rio Santa Cruz floodplain (Nabhan and Donovan 2000). Additional restoration efforts using treated sewage effluent along the Rio Santa Cruz are currently being implemented by Pima County as part of its Sonoran Desert Protection Plan—an ambitious multi-species Habitat Conservation Plan—which has strict guidelines for targeting and managing these waters to regenerate floodplain habitats for several species of conservation concern, including migratory pollinators.

In one well-documented effort, farmer-rancher Mark Larkin began to guide the "passive" ecological restoration of floodplain lands by using sewage effluent to establish riparian tree species, then seasonally reducing or increasing grazing in different patches to create healthy stands capable of long-term growth on the available water budget of treated effluent (Nabhan 2001). With his consent, the Museum staff began attempts at active restoration of pollinator habitat in 1997. While these efforts included wildflower plantings, artificial nest placements, and other pollinator population enhancement techniques described in detail elsewhere (Buchmann and Nabhan 1996; Nabhan and Donovan 2000), riparian restoration accounted for the greatest gains in pollinator abundance and diversity (Nabhan 2001). In addition to 32 species of migratory pollinators benefiting from these passive and active restoration efforts (table 1.2), we have documented some 322 species of invertebrate pollinators now in residence on Tubac Farms. There were potential seasonal population increases in other wildlife species as well. Within the past decade, ornithologists have recorded nearly 200 birds in the watershed's headwaters. Although it was not possible to assess population changes for so many species, certain Neotropical migrants show clear signs of recovery.

Building on the successes realized at Tubac Farms in the upper Rio Santa Cruz corridor, the Sonoran Institute and Center for Sustainable Environments have begun collaborations with farmers and ranchers along the San Pedro and San Simon rivers. These private-land experiments demonstrate the utility of promoting pollinators' "nectar trails" as a means to maintain wildlife corridors across private lands between protected areas. These efforts not only benefit the pollinators themselves but provide habitat for numerous other species, including habitat-modifying keystone plants and animals, frugivores, and perhaps even carnivores. The ecological restoration of riparian habitats and other wildlife habitat management efforts at Tubac Farms convince us of the value of collaborating with private-land owners. As the Wild Farm Alliance has recently proposed, we must now link their effects together to enhance the ecological functionality of an entire corridor.

Management Implications

The hypothesis that migratory pollinators are currently limited by stopover habitat quality along hyperarid portions of their corridor remains viable; efforts to protect "weak link" stopover habitats within arid stretches of these nectar trails can have benefits to the entire migratory chain. In addition, conservationists should focus more attention on remaining stopover habitats in the otherwise agriculturally dominated coastal and foothills areas of Sinaloa and Nayarit. U.S. and Mexican biologists should also con-

TABLE 1.2. Migratory Pollinators of the Upper Rio Santa Cruz Watershed of Mexico and the United States

INVERTEBRATES

Danaeus plexippus—Monarch butterfly—Mariposa monarca
Hyles lineata—White-lined sphinx moth—Palomilla
Manduca Sexta—Tobacco hornworth—Palomilla

MAMMALS

Choeronycteris mexicana—Long-tongued bats—Murcielago de lengua larga
Leptonycteris curasoae—Lesser long-nosed bats—Murcielago mescalero

BIRDS

Amazilia violicaps—Violet-crowned hummingbird—Colibrí con coronado morado
Amphispiza bilineata—Black-throated sparrow—Zacatonero garganta negra
Archilochus alexandri—Black-chinned hummingbird—Colibrí bara negra
Auriparus flaviceps—Verdin—Baloncillo
Calathorax lucifer—Lucifer's hummingbird—Colibrí Lucifer
Calypte anna—Anna's hummingbird—Colibrí cabeza roja
Calypte costae—Costa's hummingbird—Colibrí cabeza violeta
Cardinalis sinuata—Pyrrhuluxia—Cardenal pardo
Carduelis psaltria—Lesser goldfinch—Jilguero dominico
Carpodacus mexicanus—House finch—Pinzon mexicano
Colactes chrysoides—Gilded flicker—Carpintero collarejo desértico
Cynanthus latirostris—Broad-billed hummingbird—Colibrí pico ancho
Dendroica coronata—Yellow-crowned warbler—Chipa coronado
Eugenes fulgens—Magnificent hummingbird—Colibrí magnifico
Icterus cucullatus—Hooded oriole—Bolsero encapuchado
Icterus parisorum—Scott's oriole—Bolsero tunero
Lampornis clemenciae—Blue-throated hummingbird—Colibrí garaganta azul
Melanerpes uropygialis—Gila woodpecker—Carpintero del desierto
Mimus polyglottos—Northern mockingbird—Centzontle norteño
Passerina amiena—Lazuli bunting—Colorin lázuli
Phainopepla nitens—Phainopepla—Capulinero negro
Pheuticus melancephalus—Black-headed grosbeak—Picogordo tigrillo
Selaphorus platycercus—Broad-tailed hummingbird—Zumbador cola ancha
Selaphorus rufus—Rufous hummingbird—Zumbador rufo
Selaphorus sasin—Allen's hummingbird—Zumbador de Allen
Spizella breweri—Brewer's sparrow—Garrión de Brewer
Stullula calliope—Calliope hummingbird—Colibrí garganta rayada
Vermivora celata—Orange-crowned warbler—Chipe corona naranja
Vermivora virginiae—Virginia warbler—Chipe de Virginia
Zenaida asiatica—White-winged dove—Paloma ala blanca (Paloma pitayera)
Zonotrichia querula—White-crowned sparrow—Gorrión corona blanca

Source: Data from Hopp, Krebbs, Nabhan, Tewksbury, Fitzmorris, and Norman (Arizona-Sonora Desert Museum files).

sider undertaking experimental restorations of degraded stopovers historically known to have been used by migratory pollinators. In short, we must not only define corridors and determine where they are intact but also initiate restoration where they have been damaged.

Fortunately, stopover restoration efforts to ensure pollination services along corridors will likely meet with far more acceptance among farmers and ranchers than advocating for corridors to increase the movements of carnivores (Nabhan 2001). Moreover, government initiatives such as the U.S. Fish and Wildlife Service's Sonora Program, the U.S. Department of Agriculture's Wildlife Habitat Improvement Program, and the USDA's Sustainable Agriculture Research and Education Program can subsidize pollinator habitat restoration as a means to benefit both crop-yield stability and wildlife in general (Nabhan 2001). In Mexico, the Agostino Foundation, Ducks Unlimited/Mexico (DUMAC), the National Fish and Wildlife Foundation, and the National Wildlife Federation are subsidizing similar efforts to protect and restore habitats of migrants.

We do not yet know how well stepping-stone stopovers suited to migratory pollinators function for other ecological groups such as frugivores and carnivores. However, we do know that existing data on carnivore and frugivore movements will be insufficient—in and of themselves—to empirically confirm where natural corridors still function and where they are in need of restoration. In contrast, there are thousands of migratory bird, bat, and butterfly observations and flowering plant records available to empirically define nectar trails. DNA and isotope tracking techniques can empirically determine which faunal samples taken at different stopovers are from the same breeding populations. The observations about migratory pollinators' movements made by volunteer naturalists compiled on the Arizona-Sonora Desert Museum Web site (www.desertmuseum.org) may generate additional patterns regarding the location of functional corridor segments to determine which are in need of protection, restoration, or both. In addition, stopover habitats utilized by migrating pollinators capture and enhance other levels of biodiversity, such as the 322 species of non-migratory invertebrate pollinators on Tubac Farms. At the least, such collaborations between public and private sectors, and between Mexican and U.S. conservation professionals, may be the most rapid way of initiating restoration and stewardship of corridors useful for migratory (transboundary) species. It behooves all conservation biologists interested in bi- or tri-national migrants to promote private stewardship along corridors in a socially equitable and culturally sensitive manner (see chapter 3 in this volume).

ACKNOWLEDGMENTS

I thank the many researchers who contributed to this effort of the Migratory Pollinators Project and its precursor, the Forgotten Pollinators Campaign, especially those on the staffs of the Arizona-Sonora Desert Museum, the Programa para la Conservación de los Murciélagos Migratorios founded by Universidad Nacional Autonoma de Mexico and Bat Conservation International, and the University of Arizona. We particularly thank the following team members for sharing their data from the Rio Santa Cruz: Jim Donovan, Stephen Buchmann, Ty Fitzmorris, Karen Krebbs, Pete Siminski, Steve Hopp, Ginny Dalton, Keith Labnow, and Laurian Escalanti. Mark Larkin hosted us at Tubac Farms, which he manages in an exemplary manner, and Steve Walker facilitated our larger collaboration in the region. Katherine Sides kindly assisted with manuscript preparation. Support for this project was provided by the Kaplan Foundation, the Turner Foundation, the Wallace Global Fund, the Turner Endangered Species Fund, the Wallace Research Foundation, the C.S. Fund, the W. Alton Jones Foundation, the Roy Chapman Andrews Fund, and Border 21.

LITERATURE CITED

Arita, H., and K. Santos del Prado. 1999. Conservation of nectar-feeding bats in Mexico. Journal of Mammology 80(1):31–41.

Blake, J. G., G. J. Niemi, and J. Hanowski. 1995. Drought and annual variation in bird populations. Pp. 419–30 in J. M. Hagan III and D. W. Johnston, eds., Ecology and Conservation of Neotropical Migrant Landbirds. Smithsonian Institution Press, Washington, D.C.

Bowers, J. E., and M. E. Dimmitt. 1994. Flowering phenology of six woody plants in the northern Sonoran Desert. Bulletin of the Torrey Botanical Club 121(3):215–29.

Brower, L. B., and S. B. Malcolm. 1991. Animal migrations: An endangered phenomenon. American Zoologist 31:265–79.

Buchmann, S., and G. P. Nabhan. 1996. The Forgotten Pollinators. Island Press, Washington, D.C.

Bullock, S. H., and A. Solis-Magallanes. 1990. Phenology of canopy trees of a tropical deciduous forest in Mexico. Biotropica 22:23–35.

Búrquez-Montijo, A., M. Miller, and A. Martínez-Yrízar. 2002. Pp. 126–46 in B. Tellman, ed., Invasive Exotic Species of the Sonoran Desert. University of Arizona Press, Tucson.

Calder, W. A. 1993. Rufous hummingbird. Pp. 1–20 in A. Poole and F. Gill, eds., The Birds of North America, No. 53. Academy of Natural Sciences, Philadelphia, Pa.; The American Ornithologists' Union, Washington, D.C.

———. 1997. Hummingbirds in Rocky Mountain meadows. Pp. 149–68 in K. Able, ed., A Gathering of Angels: The Ecology and Conservation of Migratory birds. Cornell University Press, Ithaca, N.Y.

Carmondy, S. P. 1999. Triggers of floral anthesis of plants in the Sonoran Desert. Report to the Arizona-Sonora Desert Museum, Tucson, Ariz.

Case, T. J., and M. L. Cody. 1983. Island Biogeography of the Sea of Cortez. University of California Press, Berkeley.

Cockrum, E. L., and Y. Petryszyn. 1991. The long-nosed bat, *Leptonycteris*: An endangered species in the Southwest? Texas Tech University Museum Occasional Papers 142:1–32.

Davidowitz, G. 1998. How variable and unpredictable is desert precipitation? Ph.D. diss., University of Arizona, Tucson.

Ezcurra, E., and V. Rodrigues. 1986. Rainfall patterns in the Gran Desierto, Sonora, Mexico. Journal of Arid Environments 10:13–28.

Fleming, T. W. 2000. Pollination in columnar cacti in the Sonoran Desert. American Scientist 88(5):432–39.

Heacox, K. 1989. Fatal attraction? International Wildlife 19(3):39–43.

Heinemann, D. 1992. Resource use, energetic profitability, and behavioral decisions in migrant rufous hummingbirds. Oecologia 90:137–49.

Howell, D. J. 1974. Pollinating bats and plant communities. National Geographic Research Report 1:311–28.

Hutto, R. L. 1982. Habitat distributions of migratory landbird species in western Mexico. Pp. 211–40 in J. M. Hagan III and D. W. Johnston, eds., Ecology and Conservation of Neotropical Migrant Landbirds. Smithsonian Institution Press, Washington, D.C.

Inouye, D. W., B. Barr, K. Armitage, and K. Inouye. 2000. Climate change affects altitudinal migrations and hibernating species. Proceedings of the National Academy of Sciences 97(4):1630–33.

Lavee, D., and U. N. Safriel. 1989. The dilemma of cross-desert migrants—stop over or skip a small oasis? Journal of Arid Environments 17:69–81.

Malcolm, S. B. 1987. Monarch butterfly migration in North America: Controversy and conservation. Trends in Ecology and Evolution 2:135–38.

Miller, A. H. 1963. Desert adaptations in birds. Proceedings of the XIII International Ornithological Congress 8:666–74.

Moore, F. R., and T. R. Simons. 1992. Habitat suitability and stopover ecology of Neotropical land migrants. Pp. 345–55 in J. M. Hagan III and D. W. Johnston, eds., Ecology and Conservation of Neotropical Migrant Landbirds. Smithsonian Institution Press, Washington, D.C.

Morse, D. H. 1980. Population limitations. Pp. 437–53 in A. Keast and E. S. Morton, eds., Migrant Birds in the Neotropics. Smithsonian University Press, Washington, D.C.

Nabhan, G. P. 1999. The killing fields: Monarchs and transgenic corn. Wild Earth 9(4):49–52.

———. 2001. Nectar trails of migratory pollinators. Conservation Biology in Practice 2(1):20–27.

———. 2002. Preface. Pp. ix–xv in B. Tellman, ed., Invasive Exotic Species in the Sonoran Desert. University of Arizona Press, Tucson.

Nabhan, G. P., and A. J. Donovan. 2000. Nectar Trails for Pollinators: Designing Corridors for Conservation. Arizona-Sonora Desert Museum Technical Monograph 4. Tucson, Ariz.

Nabhan, G. P., and T. E. Fleming. 1993. Conservation of New World mutualisms. Conservation Biology 7:457–59.

Nabhan, G. P., and A. Holdsworth. 1999. State of the Sonoran Desert Biome. Wildlands Project, Tucson, Arizona.

Orians, G. H., R. G. Gates, M. A. Mares, A. Moldeneke, and J. Neff. 1977. Resource utilization systems. Pp. 164–224 in G. H. Orions and O. Solbrig, eds., Convergent Evolution in Warm Deserts. Dowden, Hutchinson and Ross, Stroudsberg, Pa.

Peterjohn, B. G., J. R. Sauer, and C. S. Robbins. 1995. Population trends from the North American breeding bird survey. Pp. 3–39 in T. E. Martin and D. M. Finch, eds., Conservation and Management of Neotropical Migratory Birds. Oxford University Press, Oxford.

Pyle, R. M. 1983. Monarch butterfly: Endangered migrants in North America. Pp. 26–42 in S. M. Wells, R. M. Pyle, and N. M. Collins, The IUCN Invertebrate Red Data Book. International Union for Conservation of Nature and Natural Resources, Gland, Switzerland.

———. 1999. Chasing the Monarchs. Houghton-Mifflin, New York.

Russell, R. W., F. L. Carpenter, M. A. Hixon, and D. A. Paton. 1994. The impact of variation in stopover habitat quality on migrant rufous hummingbirds. Conservation Biology 8(2):483–90.

Safriel, U. N., and D. Lavee. 1988. Weight changes of cross-desert migrants at an oasis—do energetic considerations alone determine the length of stopover? Oecologia 76:611–19.

Soule, M. E., and J. W. Terborgh. 1999. Continental Conservation. Island Press, Washington, D.C.

Stotz, D. F., J. W. Fitzpatrick, T. A. Parker III, and D. K. Moskovits. 1996. Neotropical Birds: Ecology and Conservation. University of Chicago Press, Chicago.

Terborgh, J. W. 1980. The conservation status of Neotropical migrants: Present and future. Pp. 21–30 in A. Keast and E. S. Morton, eds., Migrant Birds in the Neotropics. Smithsonian University Press, Washington, D.C.

———. 1989. Where Have All the Songbirds Gone? Princeton University Press, Princeton, N.J.

Waser, N. M. 1979. Pollinator availability as a determinant of flowering time in ocotillo *(Fouquieria splendens)*. Oecologia 39:165–75.

Weiss, S. B., P. M. Rich, D. D. Murphy, W. H. Calvert, and P. R. Ehrlich. 1991. Forest canopy structure at overwintering monarch butterfly sites: Measurements with hemispherical photography. Conservation Biology 5(2):165–75.

CHAPTER 2

Nectar Corridors

Migration and the Annual Cycle of Lesser Long-Nosed Bats

THEODORE H. FLEMING

Migratory behavior is a common feature in the biology of many kinds of organisms. It is particularly common in birds, where 31–42 percent of the species of north-temperate passerine birds migrate annually to the tropics (Mönkkönen et al. 1992). In North America, familiar examples of migrant birds include hummingbirds, thrushes, vireos, and warblers—species whose closest relatives live in the Neotropics. As pointed out by Levey and Stiles (1992), a disproportionate number of temperate migrants come from tropical families whose diets depend heavily on fruits or flowers or both, which live in early successional or relatively open habitats. Within these families, many tropical species undergo altitudinal migrations, which Levey and Stiles have suggested has evolutionarily predisposed these birds for undertaking long-distance migrations into the temperate zone during the warm part of the year.

The common denominator behind these altitudinal or intercontinental movements appears to be substantial spatiotemporal variation in food levels. Numerous studies indicate that food availability tends to be more patchy in open habitats (such as tropical forest canopies, forest edges, and second growth) than in tropical forest understories, and that fruits and flowers tend to be more patchily distributed than insects and vertebrates (reviewed in Levey 1988; Fleming 1992; Karr et al. 1992). Nutritional dependence on spatially and temporally variable foods, in turn, tends to select for mobility for tracking resources (Fleming 1992). Mobility and its morphological, physiological, and behavioral correlates permit species to undertake substantial seasonal migrations (Dingle 1996).

In contrast to birds, relatively few species of bats undergo long-distance migrations (Fleming and Eby 2001). Because most species of north-temperate bats migrate relatively short distances ($\leq$ 1,000 km) between summer and winter roosts, few of these species ever leave the temperate zone. Instead of migrating to the tropics, nearly all of them evade harsh winter conditions by hibernating. In North America, only five bats—two insectivores, *Lasiurus cinereus* (Vespertilionidae) and *Tadarida brasiliensis* (Molossidae), and three nectarivores, *Leptonycteris curasoae, L. nivalis*, and *Choeronycteris mexicana* (Phyllostomidae)—migrate substantial distances (up to 1,800 km) from temperate zone summer roosts into the Neotropics for the winter. Of these species, the three nectarivores barely reach the United States after migrating north from central Mexico. *L. curasaoe* and *C. mexicana* form maternity roosts in southern Arizona in the spring, whereas *L. nivalis* sometimes occupies post-maternity roosts in southwestern New Mexico and Big Bend National Park, Texas, in the summer (T. Fleming, T. Tibbetts, Y. Petryszyn, and V. Dalton, unpublished ms.).

Unlike migrant insectivores or carnivores, whose food supplies tend to be relatively uniformly distributed across habitats but are cryptic (to evade predators), migrant nectarivores (and frugivores) depend on a food supply (nectar, pollen, and fruit) that "wants to be found." These migrants cannot search widely among habitats and necessarily expect to find food. Instead, their foraging activities and migratory movements are much more tightly associated with habitats and locations occupied by their food plants. Moreover, owing to the seasonal nature of most flower and fruit supplies, they must time their migrations to coincide with the flowering or fruiting activity of their food plants. Therefore, we can envision migrant nectarivores traveling between their winter and summer habitats along "latitudinally broad paths of blooming plants pollinated primarily by migrant nectar-feeders" (Fleming et al. 1993: 72). I have called these paths "nectar corridors."

The nectar corridor concept can be applied to the lesser long-nosed bat, *L. curasoae*, because we have rather detailed knowledge about its flight and foraging behavior; it is quite straightforward to calculate the energetics involved in its migration. These calculations, for example, allow an estimate of the number of cactus flowers that are required to fuel the flight of a given number of *Leptonycteris* bats from Jalisco to the Sonoran Desert in the spring. Further, by conserving nectar corridors along the west coast of Mexico for these bats, a wide variety of other species that are partially dependent on cactus or *Agave* spp. flowers during their migrations will also benefit.

The Annual Cycle of *L. curasoae*

L. curasoae is a 23–26 g member of subfamily Glossophaginae (literally "tongue feeding") of the family Phyllostomidae (the New World leaf-nosed bats). All members of this subfamily are specialized to some extent for visiting flowers and feeding on nectar and pollen. These specializations include an elongated rostrum, a long, brush-tipped tongue, reduced number and size of teeth, and the ability to subsist on a diet of nectar and pollen (Howell 1974; Freeman 1995; Simmons and Wetterer 2002). Bats of the genus *Leptonycteris* are notable within their subfamily for their relatively large size and their long, pointed wings (Sahley et al. 1993; Simmons and Wetterer 2002). Unlike many glossophagines that live in tropical forests and are relatively slow, highly maneuverable fliers, *Leptonycteris* bats occur in relatively open arid to semiarid habitats and are built for strong, fast flight (Arita 1991; Sahley et al. 1993).

Another feature that sets *L. curasoae* apart from most other glossophagines is its roosting behavior. Whereas most tropical glossophagines live in small colonies of no more than a few hundred individuals in "cool" caves and hollow trees, *L. curasoae* typically lives in colonies containing thousands of individuals (in at least one case, more than 100,000) in "hot" caves and mines (Arends et al. 1995; Fleming and Nassar 2002). *L. curasoae* gains several physiological advantages by roosting in hot (≥ 30°C), humid caves with thousands of conspecifics as well as with large numbers of other bats. These advantages include minimal energy expenditure and evaporative water loss during the day and excellent thermal conditions for maximizing rates of embryonic development during pregnancy and for the growth of pups after they are born. Selective use of roost sites based on their microclimatic conditions comes at a price for *Leptonycteris* bats. Because it is likely that relatively few potential roost sites provide optimal physiological conditions for this species, roost sites can quickly become a major limiting factor for its populations. Protection of good roost sites is thus a major conservation issue for *L. curasoae*.

In its grossest form, the annual cycle of *L. curasoae* in western Mexico involves seasonal migrations from fall and winter roosts located in south-central Mexico to spring and summer roosts located from northern Sinaloa to southern Arizona. Not all individuals of this species undergo an annual migration (Rojas-Martinez et al. 1999). In western Mexico, females are much more likely to migrate than males, most of which appear to remain south of Sonora year-round (Cockrum 1991; Ceballos et al. 1997). As discussed below, some males migrate as far north as southeastern Arizona in the late spring.

This annual cycle is correlated with *L. curasoae*'s reproductive cycle. Unlike certain tropical glossophagine bats (such as species of *Glossophaga* and *Hylonycteris underwoodi*) that undergo two pregnancies a year (Wilson 1979), females of *L. curasoae* (and *L. nivalis*) undergo a single pregnancy each year (Ceballos et al. 1997; Fleming and Nassar 2002). In south-central Mexico, mating takes place between October and December. Observations at a sea cave near Chamela, Jalisco, indicate that thousands of males and females arrive there in September and October and, after mating, depart in December; during most of the year the cave is occupied by a few thousand males (Ceballos et al. 1997; Stoner et al. 2003). After a gestation period of about six months, females give birth to a single pup in northern maternity roosts, most of which are located in the Sonoran Desert (figure 2.1). Many females complete the final stages of their northward migration in late-term pregnancy. After the young are weaned, maternity roosts disband and adults and young bats migrate south in late summer and early fall.

Details about the timing of the northward migration and the distances that individual females migrate are still poorly known, but it appears that the dates at which females arrive at their maternity roosts varies latitudinally. Thus, females arrive at a maternity roost near Alamos in southern Sonora (27° N) by early February; near Bahía Kino, Sonora (29° N) by late March or early April; and at Organ Pipe Cactus National Monument, Arizona (32° N) by mid-April (Wilkinson and Fleming 1996; Fleming et al. 1996; Fleming et al. 1998). This variation probably reflects geographic variation in the availability of cactus flowers, which begin to bloom along the Pacific coast of Mexico in a south-to-north progression (Valiente-Banuet et al. 1996).

The late summer departure of bats from their maternity roosts is probably correlated with declining resource levels. In the Sonoran Desert, for example, *Leptonycteris* bats switch from feeding primarily on cactus flowers during the spring and early summer to feeding primarily on cactus fruits during the summer. By late summer, densities of both cactus flowers and fruits are low (figure 2.2), and bats abandon maternity roosts in

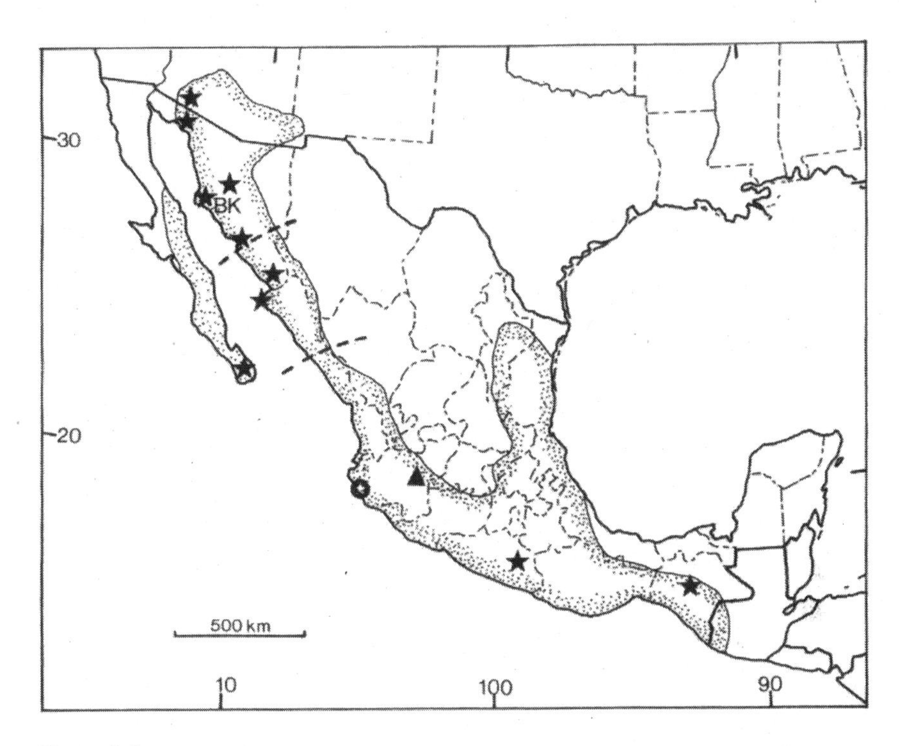

Figure 2.1. The location of major maternity roosts (stars) and non-maternity roosts (triangle) of the lesser long-nosed bat in western Mexico, especially Bahía Kino. The disk with a star indicates a mating cave in Jalisco; the semicircles represent the estimated maximum distances individuals can fly on 6 g of fat before they have to stop to refuel.

the Pinacate Biosphere Reserve, Sonora, and Organ Pipe Cactus National Monument (see figure 2.1) by mid-September in most years (V. Dalton and W. Peachey, personal communication). The maternity roost near Alamos contained very few *L. curasoae* in late October 1992 (Wilkinson and Fleming 1996), but when the bulk of the population departs from this roost is currently unknown. These observations suggest that *L. curasoae* occupies its maternity roosts for five to seven months, depending on location and availability of local plant resources, before migrating south.

Methods for Studying Migration

It is likely that nectar and pollen provide most of the fuel used by *L. curasoae* during its spring and fall migrations. Thus, knowing which flower

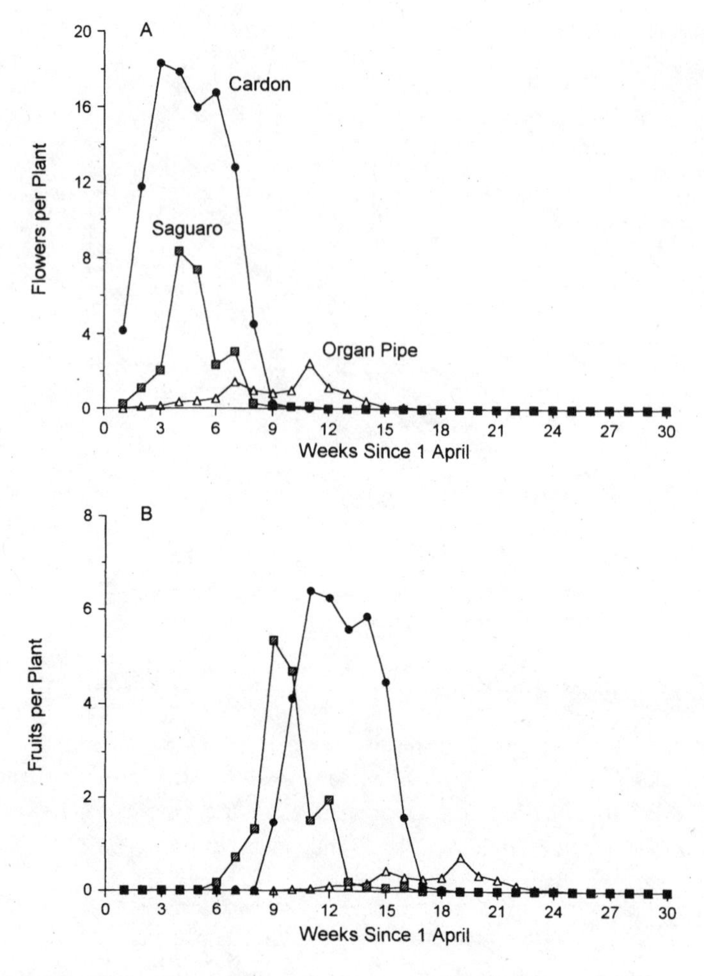

Figure 2.2. Flower (A) and fruit (B) phenologies of three species of columnar cacti, saguaro, and organ pipe near Bahía de Kino, Sonora, Mexico.

species this bat visits and their flowering phenologies and geographic distributions will help define its migratory corridors. Ideally, to study the diet of this bat, we would want detailed records of the flowers that it visits and the fruit that it eats at many locations throughout its range throughout the year. This is a tall order, and it is not surprising that we currently lack this kind of information. Data presented in Alvarez and Gonzalez (1970) and Quiroz et al. (1986) for two localities in central and southern Mexico, respectively, indicate that, in addition to visiting flowers of columnar cacti and paniculate agaves, *L. curasoae* visits flowers of *Bombax ellipticum* and *Ceiba* spp. (Bombacaceae), *Ipomoea arborescens* (Convolvulaceae), *Bauhinia ungulata* (Leguminosae), and *Crescentia alata* (Bignoniaceae). Because they are common plants that flower during the winter and spring in tropical dry forest and thorn forest along the west coast of Mexico, the species of Bombacaceae and *Ipomoea* represent important floral resources for *Leptonycteris* bats in the early phase of spring migration.

Instead of conducting detailed dietary analyses, my colleagues and I have studied the general diet of the lesser long-nosed bat at many locations throughout its range using carbon stable isotope techniques (Fleming et al. 1993; Ceballos et al. 1997). This method allows us to estimate the relative contribution of crassulacean acid metabolism (CAM) and Calvin (C3) plants to the diet of this bat because these two groups of plants differ significantly in the ratio of ^{13}C to ^{12}C in their tissues (Fleming 1995). Plants using the CAM photosynthetic pathway (for example, Cactaceae and Agavaceae) are enriched in ^{13}C relative to plants using the C3 pathway (for example, Bombacaceae and Bignoniaceae). Carbon isotope ratios are expressed in δ notation, and typical $\delta^{13}C$ values for CAM and C3 plants visited by *Leptonycteris* bats in the Sonoran Desert and elsewhere in Mexico are $-12.6‰$ and $-24.6‰$, respectively (Fleming et al. 1993).

Results of our analyses indicate that the relative importance of CAM plants in the diet of the lesser long-nosed bat varies geographically and temporally in Mexico. Bats from Baja California Sur, for example, are enriched in ^{13}C year-round, which indicates that these bats are CAM specialists. In contrast, bats living in Jalisco have low values of $\delta^{13}C$ year-round and are feeding primarily on non-succulent (C3) plants. Bats that travel north in the spring change in composition from a mixture of CAM and C3 carbon in January and February to strongly CAM by April. Bats living in the Sonoran Desert during the summer are completely CAM carbon in composition. During the fall migration, bats change from strongly CAM to a mixture of CAM and C3 carbon by November and December.

Based on these data, we proposed that *L. curasoae* uses CAM plants to fuel its spring and fall migrations (Fleming et al. 1993). In the spring, these plants include several species of columnar cacti that flower in a northward progression along the coastal lowlands of western Mexico. In the fall, these plants include several species of paniculate agaves that flower in a southward progression in the foothills and western flanks of the Sierra Madre Occidental. Based on the different flowering phenologies and geographic distributions of columnar cacti and paniculate agaves, we proposed that *Leptonycteris* bats migrate into and out of the Sonoran Desert along two different nectar corridors: a coastal cactus corridor in the spring, and an inland agave corridor in the fall.

Carbon stable isotope analysis is one of two novel techniques that we have used to study the migratory behavior of the lesser long-nosed bat. The other technique involves genetic analysis using gene sequence data from mitochondrial DNA (mtDNA). The traditional way in which bat (and bird) biologists have studied migration is to band thousands of individuals at different localities with the hope that some of those individuals will later be found or recaptured someplace else. This is a labor-intensive, inefficient method because rates of band returns are extremely low (< 1 percent), at least for bats that migrate between Mexico and the United States (e.g., Cockrum 1969). In the case of *L. curasoae*, which currently has endangered status in the United States and Mexico, large-scale banding programs are out of the question. Thus, an alternative method for determining seasonal movement patterns was needed for this species, and mtDNA analysis offers such an alternative.

In this analysis, we collected tiny samples of wing membrane from individuals from thirteen roosts throughout western Mexico and southern Arizona (Wilkinson and Fleming 1996). DNA was extracted from these tissues, and a 297 base-pair segment of "control loop" mtDNA was amplified and sequenced. For this kind of analysis, mtDNA is especially appropriate, because it evolves more rapidly than nuclear DNA (hence it provides more variation for analysis), and it is inherited maternally and undergoes no recombination (which means that, barring mutations, it is passed unchanged from generation to generation in matrilineal fashion). When we knew the mtDNA genotype (technically called a haplotype) of each individual, we looked for roost sites that contained the same haplotypes. We also subjected our genetic data to a variety of statistical analyses to look for non-random patterns of haplotype distributions.

Our analyses revealed two important patterns. First, haplotypes found in the mating cave near Chamela, Jalisco, also occurred in maternity caves in northern Sonora and southwestern Arizona as well as in southern

Baja California. This pattern provides direct evidence of seasonal movements by bats of up to 1,200 km between Jalisco and Arizona. Second, our mtDNA data suggested that lesser long-nosed bats probably use two migration paths to reach southern Arizona each year. The first path corresponds to the early spring coastal lowland route suggested by our carbon isotope data. Females mating in the Jalisco cave in October–December, for example, use this route to migrate to their maternity caves in northern Sonora and southwestern Arizona. The second path is located farther inland (along the foothills and western flank of the Sierra Madre Occidental) and brings bats from central Mexico into southeastern Arizona in the spring and summer. As mentioned above, some of the bats using this route are adult males, whereas virtually all of the bats using the coastal route are females. Why these males migrate north is currently unknown. Arizona is thus colonized by at least two genetically different waves of *L. curasoae* each year. Our data suggest that these two groups of bats reside in different parts of Mexico during the winter.

These patterns emerged from an analysis of tissue samples from only forty-nine individuals, which attests to the power of genetic analysis for discerning large-scale movement patterns. This approach has also been used recently to determine migration patterns in sea turtles (e.g., Lohmann and Lohmann 1998; Lohmann et al. 1999), birds (e.g., Joseph et al. 1999; Tiedemann 1999), whales (e.g., Baker et al. 1998a, 1998b), and humans (e.g., Murray-McIntosh et al. 1998), among other organisms.

The Energetics of Spring Migration

If some females of *L. curasoae* migrate up to 1,200 km from Jalisco to Arizona each year, how long does it take them to do this, and what is the energetic cost of such a trip? We do not yet have precise answers to these questions, but we can make some simple calculations using available data to provide some preliminary answers. Information needed to make these energetic calculations includes the amount of fat individuals deposit prior to migration, the mileage a bat can get from each gram of fat, the energetic cost of flight, typical flight speed, and the energetic value of cactus nectar (which, for simplicity, is assumed to be the primary source of fuel during spring migration). The data for making these calculations, presented in table 2.1, are

TABLE 2.1. Data Necessary for Calculating the Energetic Cost of Spring Migration in a Pregnant Female Lesser Long-Nosed Bat

Parameter	Value
Fat-free wet mass	24.5 g
Fat deposited prior to migration	ca. 3 g
Typical flight speed	40 kph
Resting metabolism[a]	2.0 ml O_2 per g-h
Cost of flight relative to resting metabolism[b]	20X
Energy from one cactus-flower visit	0.45 kJ

[a] McNab (1989).
[b] Winter and von Helversen (1998).
Sources: Cited in the text, except as noted below.

based on data from Sahley et al. (1993), Fleming et al. (1996), Ceballos et al. (1997), and Horner et al. (1998).

Stored fat is the fuel used by birds and mammals during migration, and I assume that this is also true for *L. curasoae*. Ceballos et al. (1997) reported that both males and females of this species deposit fat in the late fall. Although the precise amounts of fat deposited per individual have not yet been determined, 3 g of fat (i.e., about 13 percent of a bat's non-fat mass) is a reasonable value; some migrating and hibernating bats increase their mass by 25 percent through fat deposition in the fall (Fleming and Eby 2001). The data in table 2.1 can be plugged into the following formula (from Ewing et al. 1970) to estimate the maximum flight range that a lesser long-nosed bat can attain on a given amount of fat:

$$\text{Maximum range} = \frac{(\text{g fat})(9.4\ \text{Kcal / g fat})(\text{flight speed in kph})(10^3)}{(\text{MRr})(20)(\text{bat mass in g})(4.8\ \text{Kcal / l } O_2)}$$

where MRr = resting metabolic rate (in ml O_2/g-h). At a flight speed of 40 kph, maximum flight range for 3 g of fat is about 227 km; at a flight speed of 50 kph, it is about 283 km. These calculations suggest that individuals need to make several refueling stops to replenish their fat during a migration of 1,200 km. If they deposit 6 g of fat prior to migration and at each refueling stop, and if they fly an average of 50 kph (which is not unreasonable for these strong-flying bats), their maximum flight range would be about 550 km, and they would need to make two refueling stops between Jalisco and Arizona. Based on these calculations, lesser long-nosed bats probably stop 2–4 times to refuel during spring migration. This migration probably costs them the energetic equivalent of about 13 g of fat.

How many cactus flowers must a female bat visit to fuel her migration from Jalisco to Arizona? Thirteen grams of fat has an energetic value of 511 kJ, so a bat needs to acquire (minimally) this much energy from flowers to fuel her migration. Detailed observations of cactus flower–visiting behavior by *L. curasoae* (Horner et al. 1998) indicate that bats receive about 0.45 kJ per visit, which translates into 1,136 flower visits to obtain 511 kJ of energy. *Leptonycteris* bats make about five visits per flower, so they will need to visit about 227 cactus flowers to fuel their trip from Jalisco to Arizona. A maternity roost located just below the U.S.–Mexico border in the Pinacate Biosphere Reserve contains about 100,000 adult females, some of which probably mate in the cave near Chamela. The above calculations suggest that it will take at least *ten million cactus flowers* (assuming each flower is visited by two to three bats) to fuel the spring migration of this many bats.

How long does it take females of *L. curasoae* to migrate from Jalisco to Arizona in the spring? Again, we do not yet know the precise answer to this question, but current data and the above calculations suggest that the trip is probably slow. Many males and females leave the Chamela mating cave in December, and females do not arrive at northern maternity roosts until late March. Taken at face value, this means that the spring migration could take as long as about 3.5 months, or it could be shorter if females remain in Jalisco or elsewhere in central Mexico for some time before heading north. After they start north, they will need to stop at least twice to replenish their fat deposits. Judging by how long they remain in a transient (that is, non-maternity) roost near Bahía Kino, Sonora, before moving to their maternity roosts—about three to four weeks (Horner et al. 1998)—females could spend ≥ 1.5 months at their refueling stops en route to Arizona. This means that the trip from Jalisco to southwestern Arizona could take as long as two months in total.

Our observations near Bahía Kino, Sonora, in late March and early April (Horner et al. 1998 and unpublished data) indicate that migrant lesser long-nosed bats arrive in this area well before columnar cacti are in full bloom. A similar situation often occurs at Organ Pipe Cactus National

Monument, Arizona (T. Tibbetts, personal communication). This means that bats will have to search widely for enough cactus nectar and pollen to meet their daily energetic needs early in the cactus-flower season. Long-distance migration may be over for these bats, but they will still need to expend considerable energy on their maternity grounds finding food each night before peak cactus flowering, which occurs in late April to early May near Bahía Kino (Horner et al. 1998).

We can gain some feel for the energetic consequences of low cactus-flower density from observations my colleagues and I made at the transient roost near Bahía Kino in 1998. This roost contained about 7,000 females of *L. curasoae* when it was censused on March 26 and April 13. Based on energetic data presented in Horner et al. (1998), I estimate that these bats need to visit a total of about 140,000 cactus flowers each night to meet their daily energetic needs. Over how large an area will these bats need to search to find this many cactus flowers? In other words, what is the foraging radius of a roost of 7,000 nectar-feeding bats at this time of the year? The answer, of course, depends on flower density (the number of open cactus flowers per unit area). If flower density is high, then the foraging radius will be low, and the average bat will not have to fly very far from its day roost to find all the flowers it needs. As flower density decreases, however, the foraging radius will increase exponentially.

The relationship between foraging radius and cactus-flower density for a roost of 7,000 nectar bats is shown in figure 2.3. The Bahía Kino roost is located close to the Gulf of California, so the foraging area around this roost is a semicircle rather than a circle. Thus, I used the formula for a semicircle to solve for the foraging radius (in kilometers) that would encompass 140,000 cactus flowers given a particular flower density (number of cactus flowers per hectare). Superimposed on this curve are the actual flower densities that we measured at seven sites in the Bahía Kino area in late March to early April 1998. These densities ranged from $\leq$ 0.5 flowers per hectare within 1 km of the cave (site SK) and in the alkaline flatlands 10 km northeast of the cave (site BK) to 329 flowers per hectare at a dense stand of cardón (*Pachycereus pringlei*, the main cactus in bloom at this time of the year) 22 km east of the cave (site SN). Despite their long distance from the cave, flowers at site SN were being heavily visited by many *L. curasoae* in early April (T. Fleming and F. Molina, unpublished data). Given the low density of flowers through-

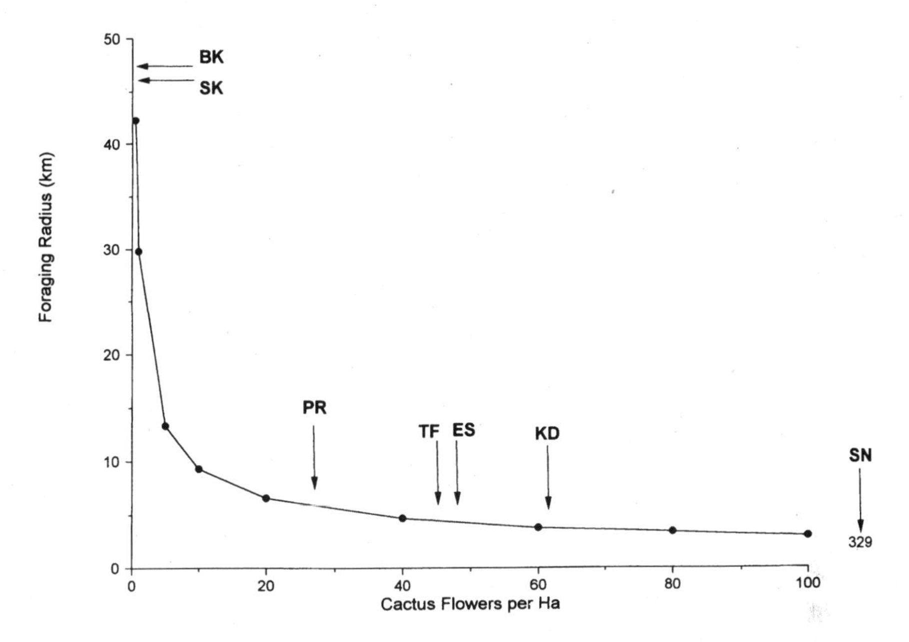

Figure 2.3. The relationship between the foraging radius of a roost containing 7,000 individuals of *Leptonycteris curasoae* and cactus-flower density at seven sites near Bahía Kino, Sonora, in late March–early April 1998. See the chapter text for further explanation of how the curve was derived.

out most of the Bahía Kino region, it seems reasonable to conclude that the foraging radius of this roost was $\geq$ 30 km at this time of the year. If this is true, then many bats had to range widely away from the roost, at considerable energetic expense, to find enough flowers each night.

Conservation Concerns

The migration of lesser long-nosed bats from south-central Mexico to the Sonoran Desert and other parts of southern Arizona can be considered an "endangered phenomenon" (see Brower and Malcolm 1991) from at least two perspectives: from that of the bats, and from that of the food plants fueling this migration. From the bat's perspective, the critical resources that need to be protected include safe roost sites and habitats containing adequate densities of food plants. Safe roost sites include caves and mines that provide protection from predators and human disturbance and that provide acceptable microclimates. Fortunately, the mating cave in Jalisco and several of the major maternity roosts in the Sonoran Desert occur on federally protected lands in Mexico and the United States and are secure. However, little is known about the locations and vulnerability of transient roosts used by lesser long-nosed bats during migration. These roosts likely are located in the states of Nayarit and Sinaloa in western Mexico, in areas that are difficult to explore because of drug activities. Ironically, the existence of *narcotraficantes* in these areas probably benefits the bats by keeping people away from their roosts.

In many areas of Mexico and elsewhere in Latin America, cave- or mine-dwelling bats are often at risk as a result of misguided vampire bat eradication programs (Villa-R 1967; Tuttle 1994; Winter and Coen 1997). A massive education program that stresses the beneficial aspects of most bats and technical training for people involved in vampire control are two obvious approaches that can help reduce the indiscriminate destruction of bats in tropical countries. Such programs are being implemented by Bat Conservation International and the Programa para la Conservación de Murciélagos Migratorios, which is a joint effort administered by BCI and the Universidad Nacional Autonoma de Mexico (Walker 1995). Even in the absence of life-threatening situations, bats differ widely in their tolerance of human intrusion into their roosts. In my experience, *L. curasoae* is quite sensitive to human presence in its roosts and is quick to temporarily abandon roosts that have been briefly visited by biologists. Unless acceptable alternative roosts occur in the area, human disturbance of roost sites occupied by *Leptonycteris* bats can have a strong negative impact on these bats, even when the disturbance is not immediately life threatening. Thus, protection of roost sites from disturbance is of utmost importance for the conservation of these bats.

Safe roost sites are only one part of the conservation equation for this bat. The other part is an adequate density and distribution of food plants, especially columnar cacti and paniculate agaves, along its migration routes. Threats to these resources include the familiar litany of wildland conversion into agricultural, ranching, recreational, and urban developments. All of these human activities are in full swing along the Pacific coast of Mexico, and the questions become: How fast are wildlands being destroyed or fragmented, and what are the consequences of these activities for migrant vertebrates, including lesser long-nosed bats? My energetic calculations suggest that nectar-feeding bats such as *L. curasoae* will be in trouble during spring migration under the following circumstances: (1) when the distance between refueling stops exceeds about 500 km; (2) when the flower density around refueling stops and other roosts forces some bats to forage more than about 50 km from their day roosts; and (3) when the absolute number of cactus flower-equivalents (the energetic value of the nectar content of one *P. pringlei* flower) drops below about 10 million units. Long distances between refueling stops or low flower densities around roosts both place inordinately high energetic demands on individual (pregnant) bats. Low absolute flower numbers along the corridor (that is, a low carrying capacity), no matter how the flowers are distributed geographically, greatly reduces the number of bats that can successfully pass along the corridor. The overall message conveyed by these calculations is that excessive fragmentation of wildlands along the Mexican coast from Jalisco to Sonora will endanger flower-dependent bats (and birds) as they migrate from their southern winter grounds to their northern summer grounds.

If *Leptonycteris* bats rely heavily on certain kinds of flowers to fuel their migrations, it is also likely that some of their food plants rely on bat visits to maximize their reproductive success, as measured by fruit and seed set. Pollinator exclusion experiments conducted in different parts of Mexico and Arizona indicate that significant geographic variation occurs in the extent to which plants with flowers that fit the "bat pollination syndrome" (see Heithaus 1982) actually depend on bats for maximum fruit set. In the Tehuacán Valley of Puebla state, for example, bat pollination accounts for 100 percent of fruit set in three species of *Neobuxbaumia* columnar cacti (Valiente-Banuet et al. 1996, 1997). In contrast, in the Gulf coast region of the Sonoran Desert near Bahía de Kino, bat pollination accounts for 25–90 percent of total fruit set in three species of columnar cacti (Fleming et al. 1996). For the geographically widespread cactus *Pachycereus pecten-arboriginum*, bat-affected fruit set is 100 percent near Chamela, Jalisco, but is only 7 percent near Alamos in southern Sonora (Molina-Freaner et al. 2003; A. Valiente-Banuet, unpublished data). Slauson (2000) conducted similar pollinator exclusion experiments with *Agave palmeri* in southeastern Arizona and found that bats (both *L. curasoae* and *C. mexicana*) accounted for only about 33 percent of total fruit set in areas near their roosts. These studies indicate that the extinction of *L. curasoae* would have a strong negative effect on plant reproductive success in some parts of its range (such as in central and southern Mexico) but not necessarily in others (such as in Sonora and Arizona). Thus, the nature of the pollinator mutualism between lesser long-nosed bats and their food plants appears to vary latitudinally. It is highly symmetrical (that is, both partners are mutually interdependent) in south-central Mexico, but it is less symmetrical (that is, bats are more dependent on plants than vice versa) in northwestern Mexico and southern Arizona. Conservation of food plants along the nectar corridors of western Mexico is absolutely essential for the evolutionary and ecological well-being of lesser long-nosed bats. But conservation of this bat is not necessarily essential for the evolutionary well-being of at least some of its food plants.

Conclusions

Bats such as *L. curasoae* that visit cactus and agave differ in many respects from other members of their subfamily. Interestingly, an independently evolved, cactus-visiting glossophagine bat, *Platalina genovensium*, which lives in the arid regions of Peru, shares many of the same morphological and behavioral characteristics found in North American cactus-visiting bats (Sahley and Baraybar 1996; Simmons and Wetterer 2002). Among these similarities is long-distance migration, presumably in response to considerable spatiotemporal variation in food availability in the Peruvian Andes. In western Mexico, many *Leptonycteris* bats migrate from tropical dry forests in Jalisco to the Sonoran Desert each spring. Most of these migrants are pregnant females, which move up a resource gradient to produce their babies in an energy-rich cactus zone. Spring-blooming columnar cacti and fall-blooming paniculate agaves are critical food sources that fuel these long-distance migrations. Destruction or fragmentation of habitats containing these plants along the west coast of Mexico will endanger the annual migration of this bat as well as many other migrant vertebrates that use cactus and agave flowers as fuel sources. Conservation of wildlands and roost sites, as well as increased public awareness of the vulnerability of bat (and bird) migration to disruption by humans, is needed to prevent the extinction of these migratory species.

ACKNOWLEDGMENTS

Many colleagues, students, and volunteers have participated in my cactus-bat studies over the years. I wish to thank them all for the ideas, energy, and enthusiasm they have contributed to this research. N. Holland and F. Molina helped me collect data in 1998. Wilfried Wolff provided useful discussion and insights into the energetics of migration. I also thank the many American and Mexican residents of Bahía Kino and elsewhere in Mexico who have helped me with my fieldwork. Special thanks go to the landowners who have allowed us to conduct research on their property near Bahía Kino. Permission to conduct research in Mexico has been kindly provided by SEDESOL and SEMARNAP. Financial support for this work has come from the National Geographic Society, the National Fish and Wildlife Foundation, the Arizona Game and Fish Department, the U.S. National Science Foundation, the Mexican CONACyT, and the Ted Turner Endangered Species Fund. Finally, our research has benefited from the Programa para Conservación de Murciélagos Migratorios, which is jointly administered by Bat Conservation International and Universidad Nacional Autonoma de Mexico.

LITERATURE CITED

Alvarez, T., and L. Gonzalez Q. 1970. Analisis polinico del contenido gastrico de murcielagos Glossophaginae de Mexico. Anales del Escuela Nacional Ciencias Biologia, Mexico 18:137–65.

Arends, A., F. J. Bonaccorso, and M. Genoud. 1995. Basal rates of metabolism of nectarivorous bats (Phyllostomidae) from a semiarid thorn forest in Venezuela. Journal of Mammalogy 76:947–56.

Arita, H. T. 1991. Spatial segregation in long-nosed bats, *Leptonycteris nivalis* and *Leptonycteris curasoae*, in Mexico. Journal of Mammalogy 79:706–14.

Baker, C. S., L. Florez-Gonzalez, B. Abernathy, H. C. Rosenbaum, R. X. Slade, J. Capella, and J. L. Bannister. 1998a. Mitochondrial DNA variation and maternal gene flow among humpback whales of the southern hemisphere. Marine Mammal Science 14:721–37.

Baker, C. S., L. Medrano-Gonzalez, J. Calambokidis, A. Perry, F. Pichler, H. Rosenbaum, J. M. Straley, J. Urban-Ramirez, M. Yamaguchi, and O. Von Ziegesar. 1998b. Population structure of nuclear and mitochondrial DNA variation among humpback whales in the North Pacific. Molecular Ecology 7:695–707.

Brower, L. P., and S. B. Malcolm. 1991. Animal migrations: Endangered phenomena. American Zoologist 31:265–76.

Ceballos, G., T. H. Fleming, C. Chavez, and J. Nassar. 1997. Population dynamics of *Leptonycteris curasoae* (Chiroptera: Phyllostomidae) in Jalisco, Mexico. Journal of Mammalogy 78:1220–30.

Cockrum, E. L. 1969. Migration in the guano bat, *Tadarida brasiliensis*. University of Kansas Museum of Natural History Miscellaneous Publication 51:303–36.

———. 1991. Seasonal distribution of northwestern populations of the long-nosed bats *Leptonycteris sanborni* Family Phyllostomidae. Anales del Instituto de Biologia Universidad Nacional Autonoma de Mexico Serie Zoologia 62:181–202.

Dingle, H. 1996. Migration: The Biology of Life on the Move. Oxford University Press, New York.

Ewing, W. G., E. H. Studier, and M. J. O'Farrell. 1970. Autumn fat deposition and gross body composition in three species of *Myotis*. Comparative Biochemistry and Physiology 36:119–29.

Fleming, T. H. 1992. How do fruit- and nectar-feeding birds and mammals track their food resources? Pp. 355–91 in M. D. Hunter, T. Ohgushi, and P. W. Price, eds., Effects of Resource Distributions and Animal-Plant Interactions. Academic Press, Orlando, Fla.

———. 1995. The use of stable isotopes to study the diets of plant-visiting bats. Pp. 99–110 in P. A. Racey, U. McDonnell, and S. Swift, eds., Bats: Ecology, Behaviour, and Evolution. Oxford University Press, Oxford.

Fleming, T. H., and P. Eby. 2001. The ecology of bat migration. Pp. 156–206 in T. Kunz and M. B. Fenton, eds., Bat Ecology. University of Chicago Press, Chicago.

Fleming, T. H., and J. Nassar. 2002. The population biology of a nectar-feeding bat, *Leptonycteris curasoae*, in Mexico and northern South America. Pp. 283–305 in T. H. Fleming and A. Valiente-Banuet, eds., Columnar Cacti and Their Mutualists: Evolution, Ecology, and Conservation. University of Arizona Press, Tucson.

Fleming, T. H., A. A. Nelson, and V. M. Dalton. 1998. Roosting behaviors of the lesser long-nosed bat, *Leptonycteris curasoae*. Journal of Mammalogy 79:147–55.

Fleming, T. H., R. A. Nunez, and L. da Silviera Lobo Sternberg. 1993. Seasonal changes in the diets of migrant and non-migrant nectarivorous bats as revealed by carbon stable isotope analysis. Oecologia 94:72–75.

Fleming, T. H., M. D. Tuttle, and M. A. Horner. 1996. Pollination biology and the relative importance of nocturnal and diurnal pollinators in three species of Sonoran Desert columnar cacti. Southwestern Naturalist 41:257–69.

Freeman, P. W. 1995. Nectarivorous feeding mechanisms in bats. Biological Journal of the Linnean Society 56:439–63.

Heithaus, E. R. 1982. Coevolution of bats and plants. Pp. 327–67 in T. H. Kunz, ed., Ecology of Bats. Plenum Press, New York.

Horner, M. A., T. H. Fleming, and C. T. Sahley. 1998. Foraging behaviour and energetics of a nectar-feeding bat *Leptonycteris curasoae* (Chiroptera: Phyllostomidae). Journal of Zoology 244:575–86.

Howell, D. J. 1974. Bats and pollen: Physiological aspects of the syndrome of chiropterophily. Comparative Biochemistry and Physiology 48:263–76.

Joseph, L., E. P. Lessa, and L. Christidis. 1999. Phylogeny and biogeography in the evolution of migration: Shorebirds of the *Charadrius* complex. Journal of Biogeography 26:329–42.

Karr, J. R., M. Dionne, and I. Schlosser. 1992. Bottom-up versus top-down regulation of vertebrate populations: Lessons from birds and fish. Pp. 243–86 in M. D. Hunter, T. Ohgushi, and P. W. Price, eds., Effects of Resource Distribution on Animal-Plant Interactions. Academic Press, Orlando, Fla.

Levey, D. J. 1988. Spatial and temporal variation in Costa Rican fruit and fruit-eating bird abundance. Ecological Monographs 58:251–69.

Levey, D. J., and F. G. Stiles. 1992. Evolutionary precursors of long-distance migration: Resource availability and movement patterns in Neotropical landbirds. American Naturalist 140:447–76.

Lohmann, K. J., J. T. Hester, and C. M. F. Lohmann. 1999. Long-distance navigation in sea turtles. Ethology, Ecology, and Evolution 11:1–23.

Lohmann, K. J., and C. M. F. Lohmann. 1998. Migratory guidance mechanisms in marine turtles. Journal of Avian Biology 29:585–96.

McNab, B. K. 1989. Temperature regulation and rate of metabolism in three Bornean bats. Journal of Mammalogy 70:153–61.

Molina-Freaner, F., A. Rojas-Martinez, T. H. Fleming, and A. Valiente-Banuet. 2003. Pollination biology of the columnar cactus *Pachycereus pecten-arboriginum* in northwestern Mexico. Journal of Arid Environments 55.

Mönkkönen, M., P. Helle, and D. Welsh. 1992. Perspectives on Palaearctic and Nearctic bird migration: Comparisons and overview of life-history and ecology of migrant passerines. Ibis 134 (supplement):7–13.

Murray-McIntosh, R. P., B. J. Scrimshaw, P. J. Hatfield, and D. Penny. 1998. Testing migration patterns and estimating founding population size in Polynesia by using human mtDNA sequences. Proceedings of the National Academy of Sciences (USA) 95:9047–52.

Quiroz, D. L., M. S. Xelhuantzi, and M. C. Zamora. 1986. Analisis palinologico del contenido gastrointestinal de los murcielagos *Glossophaga soricina* y *Leptonycteris yerbabuena* de las Grutas de Juxtlahuaca, Guerrero. Instituto Nacional de Antropologia Historia Serie Prehistoria, 1–62.

Rojas-Martinez, A., A. Valiente-Banuet, M. del Coro Arizmendi, A. Alcantara-Eguren, and H. T. Arita. 1999. Seasonal distribution of the long-nosed bat *(Leptonycteris curasoae)* in North America: Does a generalized migration pattern really exist? Journal of Biogeography 26:1065–77.

Sahley, C. T., and L. Baraybar. 1996. The natural history of the long-snouted bat, *Platalina genovensium* (Phyllostomidae: Glossophaginae), in southwestern Peru. Vida Silvestre Neotropical 5:101–9.

Sahley, C. T., M. A. Horner, and T. H. Fleming. 1993. Flight speeds and mechanical power outputs of the nectar-feeding bat *Leptonycteris curasoae* (Phyllostomidae: Glossophaginae). Journal of Mammalogy 74:594–600.

Simmons, N. B., and A. L. Wetterer. 2002. Phylogeny and convergence in cactophilic bats. Pp. 87–121 in T. H. Fleming and A. Valiente-Banuet, eds., Columnar Cacti and Their Mutualists: Evolution, Ecology, and Conservation. University of Arizona Press, Tucson.

Slauson, L. A. 2000. Pollination biology of two chiropterophilous agaves in Arizona. American Journal of Botany 87:825–36.

Stoner, K. E., K. A. O.-Salazar, R. C. R.-Fernandez, and M. Quesada. 2003. Population dynamics, reproduction, and diet of the lesser long-nosed bat *(Leptonycteris curasoae)* in Jalisco, Mexico: Implications for conservation. Biodiversity and Conservation 12:357–73.

Tiedemann, R. 1999. Seasonal changes in the breeding origin of migrating dunlins *(Calidris alpina)* as revealed by mitochondrial DNA sequencing. Journal für Ornithologie 140:319–23.

Tuttle, M. D. 1994. Saving our free-tailed bats. Bats 12(3):12.

Valiente-Banuet, A., M. del C. Arizmendi, A. Martinez-Rojas, and P. Davila. 1997. Pollination of two columnar cacti *(Neobuxbaumia mezcalaensis* and *Neobuxbaumia macrocephala)* in the Tehuacan Valley, central Mexico. American Journal of Botany 84:452–55.

Valiente-Banuet, A., M. del C. Arizmendi, A. Martinez-Rojas, and L. Dominquez-Canesco. 1996. Geographical and ecological correlates between columnar cacti and nectar-feeding bats in Mexico. Journal of Tropical Ecology 12:103–19.

Villa-R, B. 1967. Los murcielagos de Mexico. Universidad Nacional Autonoma de Mexico, Mexico City.

Walker, S. 1995. Mexico-U.S. partnership makes gains for migratory bats. Bats 13(3):3–5.

Wilkinson, G. S., and T. H. Fleming. 1996. Migration and evolution of lesser long-nosed bats, *Leptonycteris curasoae*, inferred from mitochondrial DNA. Molecular Ecology 5:329–39.

Wilson, D. E. 1979. Reproductive patterns. Part III, pp. 317–78, in R. J. Baker, J. K. Jones Jr., and D. C. Carter, eds., Biology of Bats of the New World Family Phyllostomatidae. Special Publication of the Museum of Texas Technology University, 16. Lubbock, Tex.

Winter, M., and C. Coen. 1997. Lure of the vampires. Bats 15(2):7–10.

Winter, Y., and O. von Helversen. 1998. The energy cost of flight: Do small bats fly more cheaply than birds? Journal of Comparative Physiology, B, 168:105–11.

CHAPTER 3

Conservation through Research and Education

An Example of Collaborative Integral Actions for Migratory Bats

RODRIGO A. MEDELLÍN, J. GUILLERMO TÉLLEZ, AND JOAQUÍN ARROYO

Conservation of biological diversity is one of the great challenges at the beginning of the new millennium. Only in the most recent decades have we become somewhat aware of the real dimension of the universe of life on earth, realizing that of the estimated 5 to 50 million species living on the planet today, we know fewer than 2 million. The extinction rate has been quantified and shown to be equivalent to the geologic extinction pulses such as the one that wiped out dinosaurs about 65 million years ago, except that this one is being caused by a single species, *Homo sapiens*. Ecosystems around the world are also being destroyed at alarming rates, with a severe level of erosion reducing soil productivity over thousands of square kilometers every year. The term "conservation" and its concept have become buzzwords in political, scientific, social, and economic arenas, and today even the most dictatorial governments on earth claim to be taking actions against environmental degradation and species extinction. The agenda is vast and complex, with issues such as toxic waste, water-table drop, and global warming further compounding the environmental crisis. All of these aspects are necessarily urgent, and an organized, optimized approach is the *sine qua non* to achieve success.

In 1989, after repeated informal conversations and sporadic presentations on the evident reductions in bat populations around the world—but especially in Mexico and the United States—a group of bat researchers decided to begin working toward conservation of bats in these two countries. However, we were faced with a universe of about 140 bat species that represented fully a fourth of Mexico's mammal species, with such a wide diversity of life histories, ecological traits, and habitat requirements that it was impractical and virtually impossible to focus on bat conservation as a whole.

Migration is a complex behavior that carries an organism periodically from one habitat to another (Kennedy 1985; Dingle 1995). This phenomenon exposes migratory species to many more threats than nonmigratory species. Long, seasonal movements require a finely tuned physiological mechanism that involves nutrient storage (frequently in the form of fat), precisely timed movements so that when the animals arrive to the next stop resources are available (such as reproductive partners, livable climate, and abundant food items), and preserved continuous habitat corridors that allow adequate stopovers along the way with enough resources for the population in transit, besides the winter or summer primary habitats (Berthold 1975; Rappole and Warner 1976; Brower 1985; Bairlein 1990; Gibo and McCurdy 1993).

Migration has been documented for several species of bats with extensive movements, most frequently from central Mexico to the southern United States. Some of these are included in federal lists of species at risk. For example, the two species of long-nosed bats, *Leptonycteris nivalis* and *L. curasoae*, are considered endangered in the United States and threatened in Mexico. Migration between two countries is an additional fact that makes these species relevant. Both countries involved recognize their responsibility in protecting this shared resource. Some examples of cooperative programs for migratory animals include Partners in Flight, the Monarch Butterfly Sanctuary Foundation, and Journey North.

Taking migration as the first axis for action, a binational team of researchers got together in 1990 at the National University of Mexico with the support and active participation of Bat Conservation International, and it began establishing priorities to protect bats. Migration provided a tangible, defined, and limited focus to begin working. Then, in the early 1990s, Gary McCracken and Arnulfo Moreno conducted a survey of the most important known bat caves in northern Mexico and the southern United States, the results of which confirmed our fears that bat colonies in most of that area were decreasing alarmingly and that action should be taken immediately.

A Three-Pronged Strategy

In 1994, we coalesced into the Programa para la Conservación de los Murciélagos Migratorios de México y Estados Unidos, or Program for the

Conservation of Migratory Bats of Mexico and the United States (PCMM). Our main objectives have been to prevent further damage and to recover the populations of migratory bats that move between Mexico and the United States, primarily the Mexican free-tailed bat *(Tadarida brasiliensis)* and the two species of long-nosed bats *(L. curasoae* and *L. nivalis)*. We have a three-pronged strategy to implement these objectives: research, environmental education, and conservation actions.

Research

The primary research strategy was designed to respond to several main questions: What factors intervene in the triggering of the migratory phenomenon? Is it temperature change? Onset of a reproductive event? Depletion of food resources? A combination thereof? What are the primary migratory routes followed by the three focal species? What factors affect them? What is the continuity of the corridors and wintering and summer habitats?

We use a variety of approaches to answer these questions. First, we identify particular caves that are considered priority because of the large colony in them, because of their critical geographic position within our current knowledge of the bat migration corridor, and because of the critical need of conservation owing to anthropogenic disturbances such as guano extraction, mining, vandalism, or simply being a short distance from human settlement. Then we assign one of our teams to work that cave. The coordination of these teams and standardization of dates and methodologies are among the most difficult tasks in the program.

The ideal situation is that each of the eleven working teams (in various institutions in Mexico and the United States) visits the selected priority caves (at present twenty-two caves in twelve states of Mexico; see figure 3.1) in the same week of every other month to gather the same information in the same manner so as to acquire a virtual "photograph" covering Mexico from Sonora and Tamaulipas to Chiapas. These data (table 3.1) are synthesized and condensed in the central office at the Institute of Ecology, Universidad Nacional Autónoma de Mexico.

The primary data we collect reflect the presence of all bat species in the cave and the approximate numbers of each species colony. To this end, emergences of some colonies are completely videotaped in the same man-

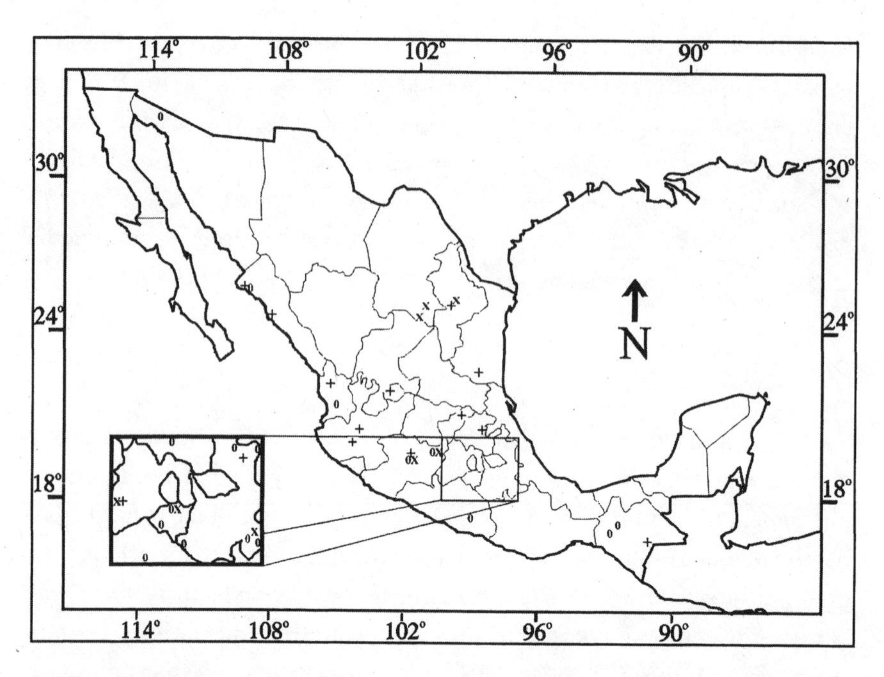

Figure 3.1. Map of Mexico showing location of the caves being monitored by PCMM personnel in twelve states of Mexico. Symbols indicate species for which data are being gathered: O = *Leptonycteris curasoae*; X = *Leptonycteris nivalis*; + = *Tadarida brasiliensis*.

ner in every visit, but most colonies are quantified visually, always by the same observers in the same manner, so that keeping a constant bias provides consistent data on intra-site seasonal population changes. We also monitor reproductive status, diet through fecal samples, pollen collected from nectarivore species' bodies, and stomach contents of a few sacrificed individuals. Samples are taken for DNA analyses, and the same sacrificed individuals are used for carbon stable isotopes to document both general diets over a long-term period and the trophic position of bats in the ecosystem. For example, a cactus and agave diet in nectar-feeding bats in the few months prior to sampling would lead to detection of high values of ^{13}C. Similarly, in the case of the insectivorous free-tailed bat, if they feed primarily on corn-eating insects such as the corn ear worm moth, the typical isotopic composition of a C3 plant like *Zea mays* will be incorporated into the moths and then to the bats that feed on them (Smith and Epstein 1971; Fleming 1995).

With this information we can also make inferences about the movements of the bats through their distribution range (Fleming et al. 1993). Molecular genetics analyses using small wing samples are useful to delineate migratory affinities and outline matriarchal lines as a source of information to learn how bat populations and lineages move in space and time. The first migration studies using molecular genetics have already been published (Wilkinson and Fleming 1996), though they are rather preliminary in their conclusions and assumptions, requiring much more detailed sampling and further analysis.

We have carried out one major workshop to coordinate data-gathering and several partial meetings to follow up on coordination and planning further steps. The associate researchers are free to carry out their own projects when gathering the data required for the PCMM. For example, projects are under way on the behavioral ecology of hairy-legged vampire bats *(Diphylla ecaudata)*, on the ecology and conservation of cave bats, and on the distribution and abundance of the bats of Mexico City.

Environmental Education

This subprogram is designed to convey the message to the general population that bats are a very useful and important component of ecosystems wherever they are found. The goal is to raise awareness in the local people living near the priority caves identified by the characteristics described above. To this end, our environmental educators have developed a series of materials and activities that can be used as modules to tailor an education program for specific regions, caves, or tasks. The materials include teacher's guidelines (for mid-elementary school grades), activity books (with games such as draw a line between the bat type and its food, circle the bugs in a cornfield, draw a bat's food, and build your own vampire bat), children's story books (fully bilingual, one for migratory insectivores, one for vampire bats, one for bats in cities, and one for pollinivorous species; three more are being developed), an itinerant exhibit, a series of fifteen-minute radio shows that won the first prize in the 1998 biannual Latin American Radio awards, and toys—one each for insectivores (a foam puzzle), vampires (a cardboard, standing bat), bats in cities (a three-dimensional puzzle), and others. Additional tools and activities include various games in the school yard, group dynamics, interpretive trails in the vicinity of the cave, talks with slides, and

TABLE 3.1. Metadata Obtained in Each Cave by the PCMM

Locality	*Leptonycteris nivalis*					*Leptonycteris curasoae*					*Tadarida brasiliensis*				
	DNA	CSI	R	P	D	DNA	CSI	R	P	D	DNA	CSI	R	P	D
El Pinacate, Sonora								X	X	X					
El Maviri, Ahome, Sinaloa											X	X	X	X	X
Ejido Juan Aldama, Sinaloa											X				
Navachiste, Ahome, Sinaloa						X	X	X	X	X					
Cueva de la Boca, Nuevo León											X	X	X	X	X
Cueva de Quintero, Tamaulipas											X	X	X	X	X
El Polvorin, Tepic, Nayarit						X	X	X	X	X					
El Salitre, Meztitlán, Hidalgo											X	X	X	X	X
Grutas de Xoxafl, Hidalgo						X		X							
Isla de Janitzio, Janitzio, Michoacán	X	X	X			X	X	X		X	X	X	X	X	X
Las Grutas, Cd. Hidalgo, Michoacán	X	X	X		X	X	X	X	X	X					
Grutas Karmidas, Zapotitlán de Mendez, Puebla						X		X	X						

Las Vegas, Tenampulco, Puebla						X		X	X						
Lenchodiego, Coxcatlan, Puebla						X		X	X						
San Gabriel, Zapotitlán Salinas, Puebla						X		X	X						
San Lorenzo, Tehuacán, Puebla	X	X	X	X	X										
Tzinacanostoc, Jolalpan, Puebla						X	X	X	X	X					
Cueva de la Peña, V. de Bravo, Estado de México	X	X	X	X	X						X				
El Diablo, Tepoztlan, Morelos	X	X	X	X	X	X									
El Idolo, Tequesquitengo, Morelos						X	X	X		X					
Cuaxilotla, Guerrero						X	X	X	X						
Grutas de Juxtlahuaca, Guerrero						X		X	X						
El Tempisque, Chiapas						X	X	X	X	X					
La Trinitaria, San Francisco, Chiapas											X	X	X	X	X
Laguitos, Tuxtla Gutierrez, Chiapas						X	X	X	X	X					

Note: Information gathered includes DNA samples (DNA), carbon stable isotope samples (CSI), reproductive information (R), population size (P), and diet (D).

question-and-answer sessions. To date, programs on insectivores, vampires, and pollinivorous species have been applied in several areas of Nuevo León, Tamaulipas, Michoacán, Jalisco, and Chiapas, and they are being expanded to all of northwestern Mexico and several other states. The oldest program, in Nuevo León's Cueva de la Boca, has been in place since 1995. Today our group is teaching the second generation of third and fourth graders. The itinerant exhibit has been on tour in various cities and towns around Mexico since 1996. The radio shows have been broadcast three times by Radio Educación in Mexico City, plus several other radio stations in various regions of the country.

Conservation Actions

Our strategy follows the principles of monitoring populations, diagnosing the negative factors that initially affected the population and were the cause of the decline, designing alternatives to counter those negative factors, and putting together a management plan for long-term conservation of the cave. Our program is primarily concerned not with implementing the management plan itself but rather with advising local non-government, academic, and government institutions on methods for implementation. We continue to monitor the caves over the long term. Under the new Mexican General Law for the Ecological Balance and Protection of the Environment, caves and other underground habitats are classified as a specific category of protected areas, called Sanctuaries. Our group sits on the committee that oversees the establishment of sanctuaries in Mexico; among the first to be declared was Cueva de la Boca, the first cave in which we began working.

Local participation is the key to achieving success. In La Boca, the participation and support of the local municipal authorities has been central to the success of the program. They have consistently shown interest and willingness to participate, even after having gone through a period of governmental change in 1997. Then, in early 1999, we began working with a non-governmental organization, Pronatura Noreste, based in Monterrey, toward establishing an agreement to work together in Cueva de la Boca. We developed a management plan for the cave, and Pronatura Noreste has been implementing it with our advice and participation. Some aspects of the environmental education work, as well as the annual population monitoring, has been carried out by us.

Interinstitutional, intersectorial, international collaboration is of critical importance for the success of this project (and, we are convinced, of many others; see Medellín 1998). Because of the dramatic shortage of scientists in Mexico, coupled with the immense biological diversity of the country and the socioeconomic crises that affect the functioning of all institutions, only by joining forces with whomever can help and has the willingness to contribute to the common end will we ever be capable of reaching the primary objective of bat conservation. The PCMM has several kinds of collaborators, including strategists and fund-raisers, policy- and decision-makers, and academic and research scientists. The primary, strategic association between Bat Conservation International and the Instituto de Ecología of the National University of Mexico is the base platform upon which the rest of the strategy rests. Funds have come from private and public sources with relative regularity, though the increasing costs of the program (a direct result of the PCMM's growth) make it increasingly hard to raise sufficient funds. Support from organizations such as the Fondo Mexicano para la Conservación de la Naturaleza and the U.S. Fish and Wildlife Service, among others, has proven to be paramount for the feasibility and short- and mid-term development of the PCMM. Both federal governments have consistently evidenced strong support for the project; bats are an increasingly frequent part of the internal agenda and of international treatises. The interaction between government and non-government institutions provides a solid foundation from which to carry out management plans and conservation agendas that are recognized by the government. Furthermore, because of the need for technology transfer and for the reasons outlined above, a binational proposal submitted to most agencies would have a more favorable reception than one that comes from a single institution or from a single country.

Of course, true collaboration is indispensable. We remember the days in which "collaborative" efforts meant taking our international colleagues to the field, showing them our best localities, helping them collect and prepare specimens, and working out their collecting and export permits, with the expectation, in exchange, of receiving a couple of mist nets and maybe being included as last author in a few papers. These pseudocollaborations, which have been typified by Mellink (1997), resulted in a benefit for only one side of the interaction. Fortunately, however, such occasions are increasingly a thing of the past. Current collaborations mean equal status in every action; intellectual input is confluent, and, as a result, collaborative papers are truly collaborative. Courses are taught in conjunction by an international team of professors to an international audience of students. Proposals are written with an enriched strategy and with a binational focus. Papers are written as a result of interactive discussions and intellectual, creative input from all individuals involved. Training has also been enhanced; some students associated with the PCMM have already been involved in training visits of several weeks to various universities abroad, expanding their horizon and significantly boosting their knowledge.

Subprojects and Metadata

After we began our activities in research, environmental education, and conservation, it became clear that this program was to become a major portion of our professional efforts. By simply doing research alone, with no formal, direct link to education or conservation, the efforts of many researchers have been lost in the depths of unfathomable scientific literature. One key characteristic of our program is the strong connection between research and education. This connection takes many different forms, from informal chats over coffee or tequilas to talks on specific topics or debates about specific scientific articles to question-and-answer sessions between the research and the education component. Popular extension articles appearing in magazines, weekly publications, or newspapers help spread the message. The fact that the program can fund specific research in addition to the information generated for the specific means of the target species is another important component of the program. We are, in effect, stimulating the formation of new human resources trained in ecology, conservation, biology, and environmental education, in a country that has a serious shortage of professionals in a field that has been rapidly turning into a priority, given the human influence on the biological diversity of the planet and the current environmental crisis. Suffice it to say that the American Society of Mammalogists has about 3,000 members in the United States, approximately ten specialists for every mammal species in that country, whereas the Mexican Society of Mammalogists has about 150 members, or about one specialist per three mammal species.

The caves in which we are working contain populations of various sizes and at least twenty-five species of bats. Two of the caves have the largest reported number of bat species in the world: Cueva de Las Vegas, Puebla, and La Gruta, Michoacán, with thirteen and fourteen reported species living there, respectively (Medellín and López-Forment 1986). The largest known maternity colony of the lesser long-nosed bat *(L. curasoae)* is being monitored in El Pinacate, Sonora, where we are working with the authorities in the Pinacate Biosphere Reserve to design and implement a protection and management plan. Our local partners involve not only local non-governmental organizations and academic institutions but also the three levels of government—federal, state, and municipal. Other projects being carried out as part of the collaborative activities for the PCMM include the following:

— We are evaluating the relationship between bat diet, phylogenetic relationships, thermoregulatory capabilities, and size, with the temperature and relative humidity of their roost in twenty caves in the state of Puebla. Metadata for this project include presence/absence and abundance of bat species in the caves, reproductive cycles of the species present, microclimatic information in all the caves and within caves in each microroost for each species, and Global Positioning System (GPS) locations of the twenty caves.

— We are analyzing the carbon stable isotope contents as an indication of the diet in the lesser long-nosed bat in space (various localities in Mexico, from Sonora and Tamaulipas south to Chiapas) and time (seasonal samples that represent phenological changes in the vegetation and their relationship with the seasonal presence of the bats in specific areas and reproduction). For this project we have data showing the presence of the lesser long-nosed bat in more than twenty localities from Sonora and Tamaulipas to Chiapas, the relative abundance of this species comparable within each site in time but not among sites, quantitative estimates of abundance in at least three caves, the diet of this species as indicated by fecal samples and sometimes pollen collected from the bats' bodies, and DNA samples to be analyzed.

— We are evaluating the effects of human visitation rates in three karstic "heat caves" (caves in which the heat is trapped and accumulated inside) in the state of Guerrero. Cacahuamilpa, Juxtlahuaca, and Cuaxilotla caves, all located in central Guerrero, fulfill the conditions to be considered heat caves. Their bat fauna was originally very similar. Today, Cacahuamilpa receives upward of 3,000 tourists per week, and it has been heavily disturbed with cement trails, illumination, garbage, latrines, and even symphonic concerts being carried out in the entrance atrium. Juxtlahuaca has a lower level of disturbance and visitation, with about 200 visitors per week. Cuaxilotla is a little farther away from any main tourist routes and has remained in a relatively pristine state. Recommendations from this study will be forwarded to the authorities in charge of these caves.

Metadata for this project include the presence, quantified abundance, and spatial distribution in space and time of each bat species in the three caves, quantification of human visitation rates to each cave, microclimatic parameters at specific sites within each cave, and reproductive information about each bat species in the caves.

— We are recording the diversity, abundance, and ecology of bats in Mexico City; up to twenty-one bat species have been recorded there. The interface between humans and bats is never more intense than in cities: bats are among the very few true wildlife groups that are capable of living in the extremely altered conditions of a large metropolis. Mexico City offers a good opportunity to learn how bats are able to adapt to living in conditions so dramatically different from nature. We are typifying the way in which these bats use resources in the city. For example, what are the common traits shared among city bat species? What are the seasonal changes in abundance?

Metadata include monthly recordings through bat detectors (Peterssen system) in several localities in the southern half of Mexico City, abundance and diversity information as indicated by mist net captures, ecomorphological information describing wing structure parameters, and body size.

— The feeding and behavioral ecology of the hairy-legged vampire bat *(D. ecaudata)* is being studied in a cave in southern

Tamaulipas that has one of the few important populations of this species, together with a large population of Mexican free-tailed bats *(T. brasiliensis)*. Information gathered includes vampire population-size fluctuations and natural history, population estimates, temporal and spatial dynamics, and seasonal changes for the Mexican free-tailed bat.

— Factors affecting species composition and microroost selection are being studied in multi-species caves such as La Gruta, Michoacán, and Las Vegas, Puebla. These two caves include in their faunas at least one of the PCMM focal species, but La Gruta has all three. Data gathered include seasonal dynamics, relative abundance, and presence/absence changes as well as reproductive information. Microclimate information at each site in the caves where particular colonies are established is also gathered, as well as the origin and topographic conformation of each cave.

In addition, we have been incorporating a database on the caves of Mexico for which biological information is available. Although we have a preliminary database of several thousand caves, there is no biological information on the vast majority. All of the caves in the PCMM are being topographically mapped and located with GPS, and a general description of the cave, its origin, and its surroundings is being entered into the dynamic database.

The Future

We anticipate being able to establish an independent office so that our personnel can devote their time and attention to the public and to coordination of bat conservation efforts. Strategy and planning for the long term have now become our top priority. We must also continue expanding our knowledge about bats in specific areas to be prepared to protect them, and we must have the facts and figures to defend them and to counter the factors negatively affecting them. In the near future, projects on the economic value of Mexican free-tailed bats, a Geographic Information System (GIS) on migratory corridors, and structured vampire bat control campaigns will begin.

By studying in detail the diet of the Mexican free-tailed bat, we will be able to put a monetary value on a per-bat, per-day basis. This will be a central tool for our continuing efforts in convincing the agricultural industry leaders of the need to participate in the conservation efforts of these bats.

Migratory corridors are still poorly understood. Although we have some bat-banding and recovery records and a limited DNA analysis, the shapes and routes of the migratory corridors are still unknown to the degree necessary for decision-making and conservation and restoration efforts. A dynamic GIS will be able to pinpoint sites and regions where conservation and research efforts should be reinforced, and in the end we hope to have a clear picture of the bats' needs to ensure the conditions necessary for the migration process. Such a GIS is being put together at Monterrey Tech in Guaymas and will definitely be necessary for future steps in the decision-making process for conservation and restoration of migration corridors.

Because vampire bats sometimes do cause real, significant damage to humans and because these humans frequently react by destroying whatever bat colonies they know of, mistakenly eliminating thousands or millions of beneficial bats and only very rarely affecting vampires, the PCMM executive committee has gotten involved in vampire bat control campaigns. Although one section of Mexico's federal government has as part of its responsibility to apply vampire bat control operations, their operational capabilities are frequently surpassed. In 2000 was the first planned vampire bat control campaign that conformed to a long-term objective by PCMM personnel. It was a landmark for our continuing growth, though we have taken care to avoid being sucked in by the strong pressure to devote all our efforts to this specific end. Eight years have passed since the start of the program. Our impact is beginning to grow, and the strategy has proved successful. Other Latin American programs have invited us to advise them, and they are beginning to achieve success, notably Bolivia, Guatemala, and Costa Rica, with three groups of bright, capable young biologists.

There is more than enough work to be carried out in the foreseeable future in the bat conservation arena in Mexico. We have taken the first steps, and now the program requires solid strategic planning, numerous participants, and a dependable collaborative scheme. Never have the circumstances and situations been more promising to achieve bat conservation; never will it be more feasible and timely than now.

ACKNOWLEDGMENTS

The PCMM would not be a reality without the active and enthusiastic participation of all the coordinators, research associates, students, and educators. At the risk of omitting some names, we would like to thank the efficient, professional involvement in the program (as researchers, strategic allies, or both) of S. Aguilar M., R. Avila F., C. B. Chávez-T., E. Clariond, A. N. Correa, G. and D. Dalton, C. Elizalde-Arellano, T. Fleming, M. L. Franco, C. Galicia, B. Gamboa, O. Gaona, C. Huerta Z., L. I. Iñiguez D., B. Keeley, T. Kunz, J. C. López V., M. Martínez C., G. McCracken, A. Moreno, G. P. Nabhan, L. Navarro, L. Reyes, A. A. Ruiz C., M. Teniente, M. D. Tuttle, S. Walker, and H. Zarza. Support for the PCMM has come primarily from the Fondo Mexicano para la Conservación de la Naturaleza, Bat Conservation International, the U.S. Fish and Wildlife Service, the National Fish and Wildlife Foundation, and the Arizona-Sonora Desert Museum, as well as varied support from Instituto Nacional de Antropología e Historia, Asociación Mexicana de Mastozoología, A. C., Pronatura Noreste, Instituto Tecnológico y de Estudios Superiores de Monterrey campus Guaymas, Universidad Autónoma Metropolitana, Universidad Autónoma de Sinaloa, Universidad de Guadalajara, ENEP Iztacala (Universidad Nacional Autonoma de Mexico), and Instituto Tecnológico de Tamaulipas.

LITERATURE CITED

Bairlein, F. 1990. Nutrition and food selection in migratory birds. Pp. 198–213 in E. Gwinner, ed., Bird Migration: Physiology and Ecophysiology. Springer-Verlag, New York.

Berthold, P. 1975. Migration: Control and metabolic physiology. Avian Biology 5:77–128.

Brower, L. P. 1985. New perspectives on the migration biology of the monarch butterfly, *Danaus plexippus* L. Pp. 748–85 in M. A. Rankin, ed., Migration: Mechanisms and Adaptive Significance. University of Texas, Austin.

Dingle, H. 1995. Migration, the Biology of Life on the Move. Oxford University Press, New York.

Fleming, T. H. 1995. The use of stable isotopes to study the diets of plant-visiting bats. The Proceedings of the Symposia of the Zoological Society of London 67:99–110.

Fleming, T. H., R. A. Nuñez, and L. da S. L. Stenberg. 1993. Nectar corridors and the diet of migrant and non-migrant nectarivorous bats as revealed by carbon stable isotope analysis. Oecologia 94:72–75.

Gibo, D. L., and J. A. McCurdy. 1993. Lipid accumulation by migrating monarch butterflies (*Danaus plexippus* L.). Canadian Journal of Zoology 71:76–82.

Kennedy, J. S. 1985. Migration, behavioral and ecological. In M. A. Rankin, ed., Migration: Mechanism and adaptative significance. Contributions on Marine Science 27 (suppl.):5–26.

Medellín, R. A. 1998. True international collaboration: Now or never. Conservation Biology 12:939–40.

Medellín, R. A., and W. López-Forment. 1986. Las cuevas: Un recurso compartido. Anales del Instituto de Biología, UNAM 56. Serie Zoología (3):1027–34.

Mellink, E. 1997. International collaboration for the collection of biological materials: Thoughts from a Mexican perspective. Proceedings of the San Diego Society of Natural History 33:38–39.

Rappole, J. H., and D. W. Warner. 1976. Relationships between behavior, physiology and weather in avian transients at migration stopover sites. Oecologia 26:193–212.

Smith, B. N., and S. Epstein. 1971. Two categories of $^{13}C/^{12}C$ ratios for higher plants. Plant Physiology 47:380–84.

Wilkinson, G. S., and T. H. Fleming. 1996. Migration and evolution of lesser long-nosed bats *Leptonycteris curasoae*, inferred from mitochondrial DNA. Molecular Ecology 5:329–39.

CHAPTER 4

Rufous and Broad-Tailed Hummingbirds

Pollination, Migration, and Population Biology

WILLIAM A. CALDER

A glittering fragment of the rainbow . . . flitting from one flower to another . . . pursuing its course and yielding new delights whenever it is seen.—J. J. Audubon

Poetic appeal evokes emotional support for nature conservation, but, for the actual implementation of conservation, nature must be approached with a scientific understanding of the functional linkages between plants, animals, and their physical environment. Ralph Waldo Emerson's (1860) insight was more appropriate: "Beyond their sensuous delight, the forms and colors of nature have a new charm for us in our perception, that not one ornament was added for ornament, but is a sign of some better health, or more excellent action."

A fine example of such excellent action is pollination, a crucial linkage in most ecosystems. Mutualisms involving pollinators have evolved many times. Many plants depend on bees; others depend on butterflies, sphinx moths, flies, beetles, honey gliders, bats, honeycreepers, honeyeaters, sunbirds, or hummingbirds. The diversity is not without some overlap, but competing alternatives provide nature with back-up security. The persistence of this pollinator diversity indicates that each is important in some special way to one or more other organisms. Each pollinator merits protection for its connections and for the ecological services provided in the process (Buchmann and Nabhan 1996; Nabhan and Buchmann 1997; Kearns et al. 1998).

In species and numbers, insects are the most abundant terrestrial animals. Even in their vast majority, insects alone do not serve in all pollinator niches, structurally, thermally, or temporally. Hummingbirds have proven that they can perform competitively; the smallest birds persist because they have an advantage. Feeding on small droplets of nectar is impractical for large animals. However, the pollinators of smallest mass—bees, flies, butterflies, flower beetles—are limited by cold and wet conditions. With homeothermic ability, hummingbirds can function as all-weather pollinators, even in subfreezing or rainy conditions—as long as they have access to a fuel supply. Many northern temperate plants benefit from, or even depend on, hummingbirds to pollinate them.

Hummingbird evolution has left no fossil record, but hummingbirds now range from Neotropical to neotemperate realms. Several species in the genera *Archilochus, Calypte, Selasphorus*, and *Stellula* migrate annually between tropical and temperate regions. Their systematics are often revised, but the current count of 328 species qualifies the Trochilidae as the most species-rich, non-passerine bird family—the second largest bird family in the Western Hemisphere, and the most highly specialized nectar-feeding, pollinating birds (Sibley and Monroe 1990; Johnsgard 1997; Schuchmann 1999).

Some of these species face greater thermogenic demands than their tropical ancestors, by claiming the higher latitudes and elevations where floral resources are available only seasonally. Their ability to enter hypothermic torpor can conserve energy for short periods in some situations, but the limited energy reserves, with the birds' small size, cannot be stretched for a season's duration (Calder 1994). Consequently, migration is the only solution to the problem of meeting energy requirements and surviving year-round.

In his theory of the evolution of migration, Cox (1985) regarded the Mexican Plateau region as "a 'staging area' for the evolution of many patterns of disjunct migration." Migratory behavior has certainly been fundamental to the range expansion and adaptation of *Selasphorus* hummingbirds. *Selasphorus* is one of fifteen genera of hummingbirds found only north of Panama and therefore assumed to have come from a secondary radiation of hummingbirds in North America (Mayr 1964; Johnsgard 1997).

With the final retreat of Pleistocene continental glaciation, secondary radiation provided species diversity, and primary plant succession produced new habitat. This set the stage for *Selasphorus* to colonize the Pacific Northwest and the Rocky Mountains. Eventually the rufous hummingbird

(S. rufus) extended its range to Prince William Sound, Alaska—far from its Neotropical origins and its radiation onto the Mexican Plateau (Calder 1987, 1993). Meanwhile, the broad-tailed hummingbird *(S. platycercus)* pushed its range limits north from the Mexican highlands into the southern and central Rocky Mountains and the eastern Sierra Nevada, to breed as far north as Utah, Wyoming, and southeastern Idaho. It had been thought that the broad-tail "may be more closely related to *Archilochus [colubris]* than to other North American species of *Selasphorus*" (such as the rufous and Allen's hummingbird *[S. sasin]* superspecies; Mayr and Short 1970). However, DNA analysis (W. H. Baltosser, personal communication) supports an alternative scenario, that the broad-tail diverged from common ancestry with this rufous and Allen's superspecies as their respective spring migrations evolved.

These two species subdivided further by migration patterns (rufous) and timing of molt (broad-tail). The U.S. broad-tail (BTUSA) expanded to colonize high montane habitat from Arizona and New Mexico, up the Rockies to the greater Yellowstone ecosystem. The BTUSA migrates to winter in Mexico with the Mexican broad-tail (BTMEX), but its annual cycles differ. The BTMEX molts in early summer and breeds from September to December, whereas the BTUSA molts in mid-winter and breeds from April or May through July (Calder and Calder 1992).

All rufous hummingbirds migrate north in late winter and early spring, through the Pacific coastal states, and then diverge into Pacific Northwest breeders (RUPNW) and northern Rocky Mountain breeders (RURMt). The RURMt may have diverged from RUPNW stock, expanding inland to Idaho, western Montana, British Columbia, and Alberta. Later, they use two flyways to return to central and western Mexico for the northern winter: the coastal ranges and Sierra Nevada route, and the Rocky Mountain flyway. The simplest assumption is that these are, respectively, for RUPNW and RURMt, but banding recovery records indicate that some RUPNW use the Rocky Mountain flyway when migrating southward.

When southbound rufous migrants began to use the Rocky Mountain flyway, reproductive isolation from the BTUSA was probably maintained by post-reproduction hormonal changes in rufous hummingbirds (Immelmann 1971), made possible by the milder Pacific climate influence on the Pacific states and northern Rockies, which allowed earlier spring migration and breeding seasons. The rufous is the champion high-latitude hummingbird. Is it an indicator or keystone migrant pollinator? If so, what is its conservation status?

Their Role in Pollination

Rufous hummingbirds have been observed to feed from several species with typical coevolved, red tubular flowers, but at the northern limit of rufous distribution in coastal southern Alaska, Grant and Grant (1968) found "only five plant species with floral characters suggesting hummingbird pollination." Throughout their annual range, the rufous use several non-hummingbird species that are open-cupped and upward-facing. Several of these appear to have begun a coevolutionary path or process, acquiring small tubular and/or pinkish or magenta corolla (table 4.1).

The intense metabolism of hummingbirds (ca. 8 kJ per g per day) supports their high body temperatures. This enables these small birds to fly in chilly weather—in the mountains at low or mid-elevations, or north to Alaska and south to Tierra del Fuego—where and when the insects are mostly grounded. This high metabolism also creates high demand for nectar sugar, which ensures hummingbird visitation when pollination is needed to a fairly precise level of foraging demand and pollinating potential. For example, the territories of migrating rufous hummingbirds included 1,595–3,961 flowers of *Castilleja linariaefolia*. When floral density was manipulated to half, the time spent foraging increased from 21 to 26 percent of the day, and territory size increased from 74 to 450 percent in compensation (Hixon et al. 1983).

Befitting the Emerson quote above, hummingbirds and flowers do more than just look pretty. Forests are crucial to watershed protection and control of runoff, such as would result from the heavy precipitation in the Northwest, and stability of mountain slopes throughout hummingbird flyways and winter habitats. Devegetation, by whatever means, leaves slopes without the protection of branches and organic debris that would normally slow or retain rainwater. Secondary plant succession restores vegetative protection after devegetation by fires, clear-cut logging, road-building, slope failure, wind storms, and heavy precipitation. Devegetation can also be the result, indirectly, of the increased intensity or frequency of El Niño or La

Table 4.1. Flowering Species from which Rufous Hummingbirds Have Been Observed to Feed and Can Be Tentatively Assumed to Pollinate

**	Red columbine (*Aquilegia formosa*)
**	Red and orange paintbrush (*Castilleja* spp.)
**	Honeysuckle (*Lonicera* spp.)
*	Beardtongue (*Penstemon* spp.)
*	Currant (*Ribes sanguineum*)
*	Salmonberry (*Rubus spectabilis*)
*	Fireweed (*Epilobium angustifolium*)
	Lily (*Erythronium grandiflorum, Lilium columbianum*)
	Huckleberry (*Vaccinium ovatum* and other *V.* spp.)
	Menziesia (*Menziesia ferruginea*)
	Pacific madrone (*Arbutus menziesii*)
	Snapdragon (*Scrophularia montana*)
IN MIGRATIONS (CALDER 1993; VAN DEVENDER ET AL. 2000)	
**	Chuparrosa (*Justicia californica*)
**	Ocotillos (*Fouquieria macdougalii, F. splendens*)
**	Scarlet gilia (*Ipomopsis aggregata*)
**	Paintbrush (*Castilleja* spp.)
**	Cigarrito (*Bouvardia ternifolia*)
**	Texas betony (*Stachys coccinea*)
**	Red beardtongue (*Penstemon barbatus*)
*	Pink beardtongue (*Penstemon parryi, P. pseudospectabilis*)
*	Fireweed (*Epilobium angustifolium*)
	Purple larkspur (*Delphinium barbeyi, D. geranioides*)
	Bee-balm (*Monarda fistulosa*)
	Butter and eggs (*Linaria vulgaris*) (introduced, Eurasian)
	Snapdragon (*Scrophularia montana*)
	Bee plant (*Cleome serrulata*)
**	Limita (*Anisacanthus andersonii, A. thurberi*)
	Morning glories (*Ipomoea arborescens, I. bracteata,* **I. coccinea*)
	Palo chino (*Havardia mexicana*)
	Pineapple sage (*Salvia elegans*)
WINTER RANGE (DES GRANGES 1979)	
*	Sage (*Salvia iodantha, S. mexicana*)
	Currant (*Ribes ciliatum*)
	Bush groundsel (*Senecio angustifolius*)
	Night-blooming jessamine (*Cestrum terminale*)
	Teposoma (*Buddleja cordata*)
	Tree morning glory (*Ipomea* spp.)

**Apparently coevolved red to red-orange, tubular.

*Apparently influenced, having small tubular and/or pinkish or magenta corolla.

Niña—that is, El Niño Southern Oscillation cycles and violent storms attributed by some scientists to the greater atmospheric energy of global warming.

Many of the plants in the secondary succession are chosen for foraging by hummingbirds. Ecologists have tended to pick the more obviously coevolved hummingbird flowers, with red, long, tubular and pendulous flowers, for pollination studies. Many coevolved plant-pollinator pairs are fairly specific (such as hummingbirds with red *Salvia* spp., *Justicia californica, Aquilegia formosa, Ipomopsis aggegata*, or *Penstemon barbatus*).

As nectar or pollen availability varies, these hummingbirds may cross over to exploit the flowers typically used by other kinds of pollinators. Rufous hummingbirds are attracted to some shallow, open flowers—presented face-up to a pollinator—and often in colors other than red; to be nutritionally important in their open forest habitats, the flowers have been undergoing secondary succession. They also visit Indian paintbrushes *(Castilleja)* with inconspicuous flowers surrounded by colored bracts (not petals). Grant and Grant (1968) thought these were originally hummingbird pollinated, and some later spun off yellow and purple forms that opportunistically attracted insects. Thus, coevolution is a two-way street.

It remains to be determined how effective the broad-tailed and rufous hummingbirds are as pollinators of flowers that previously evolved to be insect-pollinated. The pink *Penstemon pseudospectabilis* of Arizona and Sonora attracts broad-tailed, broad-billed *(Cynanthus latirostris)*, and black-chinned *(Archilochus alexandri)* hummingbirds, and a migrating transient rufous will defend patches as temporary territories. However, *P. pseudospectabilis* retains a corolla wide enough for bees to enter and pollinate. This combination of bees and hummingbirds is referred to as a mixed pollination system. The amount of seed-set was doubled when only small halyctid bees had access to the flowers (Lange and Scott 1999).

Further research on multiple-pollinator effectiveness should be a research priority, given the practical significance of problems caused by deforestation throughout the migratory range of rufous hummingbirds. If they do pollinate these plants effectively, then hummingbirds would expedite the natural restoration of vegetation with significant ecological influence, indirectly providing berries for birds and bears and protecting salmon spawning beds. Furthermore, if climatic disruption denies a nectar supply for migratory refueling, then hummingbirds may not return to serve this web.

Current Status

Phillips (1975) cautioned that "Accurately determining the migration routes and dates for each species (including sex- and age-classes) will furnish merely a framework on which to build an understanding of the year-round activities of these amazing birds and to make interspecific comparisons." Data from the National Audubon Society's Breeding Bird Survey suggest possible declines in some rufous hummingbird populations (Peterjohn et al. 1995; Muehter 1999). These data remain suggestive, not definitive; clearly, we need more knowledge about migratory hummingbird demography and vulnerability (Russell et al. 1994).

The rufous hummingbird received a "moderate priority" score of twenty (thirty was the highest priority) on the National Audubon Society's WatchList of bird species for which there are indications of possible population declines (Muehter 1999). If these simply reflect a reduction in flower abundance as plant succession proceeded along traditional routes (according to the Breeding Bird Survey), there may be no cause for alarm. In either case, background information is needed to signal early warnings. Here is what we know thus far:

- Survivorship data exist for only one population of one species—the broad-tailed hummingbird (Calder 1990).
- Population density has been estimated for only two species—the calliope *(Stellula calliope)* and broad-tailed hummingbirds (Calder and Calder 1992, 1994, 1995).
- Most relative abundance data come from breeding bird surveys. These are difficult to evaluate because of the confounding effects of secondary plant succession. Rufous hummingbirds may be seen most abundantly in recently logged or burned forests, feeding on the flowering perennials and shrubs most abundant after disturbance. As succession proceeds and the forest closes in, there are naturally fewer flowers that would attract and support hummingbirds.
- Information about migration routes within broad nectar corridors is meager. Hummingbirds can escape winter cold and unavailability of food by migrating, but the distance traveled between

refueling stops, which are crucial for reaching winter habitat, is limited by the down-scaling of airspeed and endurance to match their small body sizes. Flight range is roughly proportional to one-quarter the power of lean body mass ($M^{1/4}$; calculated from Tucker 1974). For a rufous hummingbird, 2.5 g of fat might sustain a flight distance of about 1,095 km (650 miles) with no wind. Thus, a 4,320 km breeding-to-wintering rufous migration from southern Alaska to central Mexico might require four or more flights plus stopovers, in different habitats, for refueling, or even more if headwinds are encountered or if nectar is in short supply at any stop, which would limit potential fattening. A transient normally takes one to two weeks to replace the protein and fat metabolized on a completed segment. This extends the migration to four to eight weeks—a minimum of a month en route. When nectar availability is poor in drought years, the trip and its refueling stops would be prolonged. If there is no accessible food waiting at a stopover site, life may be terminated.

Actual records of migration routes taken by individuals are sparse. The rufous seems to be the most abundant species during migration. Unfortunately, only a dozen interstate recoveries have been made of banded rufous, and only one of these occurred less than a month after banding (Calder 1999). That southbound female rufous might have added the usual 2 g of fat to her 3.5 g body mass after capture in Montana before departing. This would have taken at least a week before flying 1,202 km to the point of recapture, fifteen days later in Gothic, southwest Colorado. After her arrival in Gothic, the circumstantial evidence of body mass suggests that she had added 0.2 g (about four-fifths of a typical day's mass increase) of sugar solution water in her crop and gut, protein in muscle, or subcutaneous fat (Calder and Jones 1989). Whether the 1,202 km from Montana to Colorado had been spanned in one or two flights is open to speculation.

Habitat Loss and Degradation

Of six criteria for estimating the conservation priority in the Audubon WatchList (Muehter 1999), two are demographic: relative abundance and population trend. The other four are geographic: breeding and winter distributions, and threats to breeding range and non-breeding range. Omitted from specific consideration is migratory stopover habitat. For the migrants, this is analogous to flying an airplane when fuel is running low, the radios are inoperative, and the pilot is clueless about any refueling facility within range. The threats from habitat degradation and climate variation are especially critical in drought years, when the flight path must cross a vast expanse, say 800 km, of the most hyperarid stretches of Sonoran Desert or, alternatively, the Sea of Cortés (see chapter 1 in this volume).

Flowers suitable for hummingbirds are not widely distributed in these areas, but instead are concentrated in a few small or narrow areas along some arroyos and on some slopes where there is minimal subsurface moisture to go into nectars. Massive drives to convert native habitat to cattle forage on alien grasses threaten the availability of nectar-producing flowers used by migrating hummingbirds. In particular, African buffelgrass *(Pennisetum ciliare)* is more fire resistant underground than the native flora, but it is highly flammable aboveground, where the heat can kill native plants. Thus, wildfire can eliminate the native plants without cutting back on the alien grasses (Búrquez et al. 1996).

Responses to Climatic Variability

Climate has multiple influences on individual survival and on migration and reproduction success. One needs only to witness the plight of a migrant stranded at a desert stopover in a drought year to appreciate the predicament (Miller 1963; Calder 1999). Indications of population declines may reflect climate-related vulnerability during migration and rest stops. A rufous-sized hummingbird traveling 1,095 km on one leg of a migration might metabolize 2 g or more of fat in less than 24 hours. That is about three times the energy usually burned in a 24-hour day of metabolism for temperature and fluid regulation, perching, feeding, and the male's territorial and courtship behavior or female's incubating, brooding, defending, and feeding her offspring. Hummingbird survival therefore depends on the ability to meet fuel needs for the highest energetic level in the annual cycle, on a specialized diet.

Little is known about dispersal strategies of regional populations of rufous hummingbirds in seeking out newly opened forest habitats. Although confirmation is needed, tentative indications of declines in the rufous populations drive home the need to know more about the biology of these special pollinators. Such knowledge is needed to evaluate early warnings versus false alarms and to learn the causes and consequences of real declines. We would want to know whether the declines were due to loss of breeding habitat, loss of wintering habitat, or loss of migration stopover habitat or resources, and why and where these losses occurred. For example, drought may be particularly disastrous where hummingbirds must cross deserts, which offer only limited opportunities for refueling. Changes in wind patterns or other climate effects may impair orientation or increase flight costs if tailwinds are not available.

Population and Reproductive Biology

Hummingbirds are apparently at the lower size limit for the avian egg plan; they seem unable to incubate more than two eggs at one time. In the short seasons at high elevations and latitudes, there is only time for a single five-to-six-week nesting cycle, with only 46 percent nesting success (Calder et al. 1983); a female broad-tail must live at least two years to achieve replacement. It would pay to avoid uncertainty in the second nesting year and return to the same location. Hence they show breeding-site fidelity.

A specialized diet dependent on drought and cold-sensitive plants and the consequences of small size make hummingbirds dramatically vulnerable. For example, hurricanes can directly destroy flower and fruit resources and thus decimate populations of their pollinators indirectly. After Hurricane Lili in 1996, bananaquits *(Coereba flaviola)* and Bahama woodstars *(Calliphlox evelynae)* were absent from a study site (Rathcke 2000). To understand the population biology of continental long-distance migrant species, we must know how much hummingbird abundance varies. Given their low reproductive capacity (for broad-tails: two eggs, one clutch, 46 percent nesting success, two-year female life expectancy), how long does it take for recovery from a population drop caused by climatic, migration, or other disasters? How much of the recovery is due to meta-population source-sink biology? How does this vary between habitats?

As Bailey and Niedrach (1965) reported, "the winter of 1957–1958 was a disastrous one for hummingbirds, for less than one-fourth of the 1957 populations of broad-tails appeared in 1958 and 1959 [in the Daniels Park area south of Denver], and similar reports were received from observers in other parts of the state." This regional population decline did not, however, permanently threaten broad-tail species survival. Nevertheless, it did show the vulnerability of migratory species to freak weather and climatic variations, which could be part of anthropogenic climate change.

Three decades later, David Inouye, Nick Waser, and I noted a similar pattern in the breeding population around the Rocky Mountain Biological Laboratory in Gothic, Colorado. We had been banding this population with varying intensity since 1972 (Calder et al. 1983; Calder 1985) while studying pollination, behavior, osmotic and temperature regulation, microclimatic adaptation, and population biology. Most captures were with hand-operated traps containing feeders with sucrose solutions, which the birds sought intensely early in the season until meadows filled with flowers (May to mid-June), when feeders became relatively unattractive. By mid-July, because of feeding chicks and competition with transient southward-migrating rufous hummingbirds, energy demands increased and hummingbirds were drawn back to the feeders for easy capture. Adult birds that had evaded capture earlier were commonly added to the summer census between mid-July and mid-August, when departures of male broad-tails for Mexico were progressively more frequent.

Following the unusually late and cold spring of 1995, the population declined an estimated 57 percent in one year. Compared with years of normal onset of male territoriality and female nesting in late May, the summer of 1995 came late, in association with slow melting of a record snowpack. Waser and Inouye observed few broad-tails before their banding began on June 13. A July 3 snowstorm further impeded breeding. Dead hummingbirds were reported all the way down to Almont (ca. 2,440 m elevation). The study of rufous hummingbirds in Washington delayed our arrival in Gothic until July 8. Early captures by Waser and Inouye, plus our captures later that season (July 8–August 8) amounted to only 118 birds (36 percent of the mean annual capture number for 1987–91, and 43 percent as many adults as the 277 we had processed in 1994). Breeding late in the short season, with low reproductive capacity at best, apparently led to negligible increases in potential parents for 1996, when only 110 adults were captured. However, this was followed by an increase to 130 adults in 1997. The Gothic broad-tail population was not monitored in 1998, while we were focusing our efforts on the rufous hummingbird in Washington, then during the southward migration in New Mexico. During the hiatus, a female broad-tail banded in Gothic in 1994 was recaptured in 1998, east of the Continental Divide near Salida, Colorado, convenient to where Brenda Wiard was banding (personal communication). This serves to remind us that local disappearance, even from a population with high site fidelity, does not necessarily mean mortality.

The Gothic population continued to grow: to 218 adults in 1999, and 324 by early August 2000. This recovery of population size took five years. How much of this was from local reproduction and how much was from source populations elsewhere is unknown. However, the rates of increase (18 percent from 1996 to 1997, 26 percent per year from 1997 to 1999, and 49 percent from 1999 to 2000) considerably exceeded the reproductive potential from local breeding alone. Thus, it appears that this higher elevation's population barely replaces itself in good years, becoming a textbook "sink population" for emigrants—probably from lower elevations—after crashes due to environmental stochasticity.

Finlay (1999) reported a drop of 75 percent in female rufous and 48 percent in males at his three banding sites on Vancouver Island, British Columbia, in 1998. Corrected to birds per hour of trapping (both sexes), the drop was 60 percent. Our captures in Stabler, Washington, for the first week in June declined from thirty-three in 1997 to nine in 1998 (five of the nine were recaptured, wearing 1997 bands). However, larger samples over longer periods, temperatures, and phenological data would be necessary to conclude anything about the 1998 "decrease," especially considering the numbers captured in the next two years (thirty-one adult birds in 1999, but earlier in the year, May 14–16; seventy-four adults banded in 2000 and eleven recaptured with bands from past years, in eight days, June 12–15 and June 25–28). Hence a banding program over the season, such as we have had with broad-tails at the Rocky Mountain Biological Laboratory, is necessary to span early versus late seasons, and a range in flower availability.

Possible Weak Links and Perilous Passages

Hummingbird survival depends on meeting the fuel needs of an intense metabolism with a specialized diet, which most northern communities are unable to provide on a year-round basis—they have neither the nectar nor appropriately small insects for proteins. The evolution of migration was a major factor in the latitudinal success of this radiation, given the temperate seasonality of food resources, allowing hummingbirds to retreat from winter cold and unavailability of food. Hummingbirds that breed in higher latitudes go to lower latitudes for the winter, and hummingbirds that breed in mountains at lower latitudes descend to lower elevations for the off season.

In general, the smaller the bird, the smaller its fuel (fat) supply and the shorter its distance endurance on that fuel. Therefore, hummingbirds must break long migrations into feasible segments with refueling stops en route.

Returning annually to the highest latitude attained by hummingbirds, the rufous makes the longest hummingbird migration. In proportion to size (ca. 48,600,000 body lengths), this is also the longest bird migration. Broad-tailed hummingbirds migrate from western Mexico to the central Rocky Mountains and Sierra Nevada of eastern California and Nevada. Broad-tails that are not yet through with their breeding season do provide seasonal competition for the same resources being sought by rufous hummingbirds arriving in a southbound migration. Between breeding and winter habitats while southbound in July, rufous hummingbirds crowd flower patches in the Sierra Nevada (presumed to be RUPNW breeders) and the northern Rocky Mountains (probably RURMt breeders). Paradoxically, of the few banded rufous intercepted in the Colorado cross-over, RUPNW outnumber RURMt (Calder 1993, 1999).

Broad-tailed hummingbirds breed in mostly higher montane habitats, May to July, in the central and southern Rockies (BTUSA). They molt in Mexico during early winter. Others in the eastern Mexican highlands (BTMEX) breed in fall and molt in late spring (Calder and Calder 1992). Actual routes and routing strategies are known only in vague terms, based on where and when they have been seen or collected, but it seems reasonable that vulnerability during migration might cause apparent or future population declines. Too few hummingbirds have been banded and subsequently retaken to yield a database adequate for analysis; we have only a series of preliminary pieces of the overall puzzle (Calder 1999).

Of the two major flyways used on either side of the Great Basin Desert by RUPNW and RURMt, the simplest expectation is that RUPNW would follow the Coast Ranges and Sierra Nevada, and the RURMt would follow the Rocky Mountain Cordillera to the Sierra Madre Oriental (Phillips 1975; Calder 1993). However, eight of eleven interstate recoveries of banded rufous were using the Rocky Mountain flyway to travel southeasterly (Calder, 1993, 1999, unpublished research). Generalizing from so few data is stretching things quite a bit, but one wonders whether these are typical birds or are unusual birds like the exponential increase in rufous being banded in winter at feeders in the Southeast (Hill et al. 1998). Migration is not a given; it must evolve to adjust to temporal and spatial changes in climate, part of a species' adaptations to the physical environment. Patterns of distribution and wintering are changing, and the number of scientists who offer evidence of global climate change is increasing. What is actually happening can be known only from careful monitoring.

Missing Information

Knowledge about routing, timing, the energetics of hummingbird migrations, and the population biology of hummingbirds is fundamental to ensuring persistence and reliability in pollinating services. The four kinds of information needed are listed here with possible approaches to getting this information:

1. Variation in abundance (population size) from year to year. Approach: monitor the abundance of the primary-focus pollinator species (rufous) in nature via mist-netting, banding, and transect (field) counts. Use counts at feeders for quick-and-dirty tracking of geographic routes and migration periods.

2. The "tightness" or species-specificity of mutualisms, between plants dependent on transient migrants for pollination and the hummingbirds that rely on plant nectar for refueling during migration versus redundancies (alternate pollinators, alternate nectar sources) during migration versus breeding seasons. Approach: study the abundance of other members of the pollinator community (a) as they compete for nectar resources, and (b) as they serve as backup pollinators for stochastic events or as redundant pollinators in good years. This is significant for corridor identification as well as for the ecological services of pollination. This involves observation of foraging, identification of pollen adhering to hummingbirds and matching it to

reference pollen collections, relating this to plant distributions, and perhaps experimental manipulation.

For instance, in Chamela, Jalisco, Gryj et al. (1990) found that *Combretum fruticosum* was incidentally visited by hummingbirds but effectively pollinated more by passerines, such as *Vermivora ruficapilla*. Its congeners *V. cellata* (orange-crowned warblers) and *V. luciae* (Lucy's warbler) foraged in patches of hummingbird flowers in northern Sonora in the spring migration of 2000.

3. The nature of migration routes of rufous, broad-tailed, and Costa's *(Calypte costae)* hummingbirds. Approach: identify flight origins (regions of birth) with molecular techniques. Banding anecdotally yields tantalizing suggestions, but staking hopes for route information on banding recoveries—the traditional source of route information—is more an exercise in wishful thinking than an analysis of actual recaptures. Where migrants cross hyperarid lands, we should seek to identify flyways (broad, diffuse north-south streams). Or, if birds are more narrowly confined to following ridges, canyons, and riparian areas, we should focus on the habitat quality of entire landscapes or homogeneous vegetation corridors.

My colleagues at the University of Arizona are now seeking informative microsatellite DNA loci to identify patterns to test the connection between sample migrant populations with natal regions. Completion of this DNA reference-pattern collection will be the basis for any future understanding of migration monitoring data, because it would enable us to read geographic genetic patterns like we read band numbers. A reference database from "normal" migrations could subsequently allow us to decide whether decreases in abundance were from Pacific Northwest or from northern Rocky Mountain rufous populations. An even drop would suggest impacts beyond the breeding habitats.

4. The effects of climate on migration, both direct and via nectar and insect availability, as they relate to physical conditions accompanying cycles such as El Niño and La Niña, or anthropogenic climate change. Approach: incorporate analysis of weather and monitoring migration-related effects and responses such as (a) energy and nutrient availability as they affect pre-migratory preparations such as molt completion and fuel storage, (b) stopovers for refueling en route, (c) habitat conditions such as shelter,

food, and bird crowding and interaction, and (d) nectar and insect supplies affected by temperature, available water, favorable winds, annual patterns, El Niño, La Niña, and other climate trends.

Research and Monitoring Plans for the Future

Although recaptures of banded hummingbirds are relatively infrequent, our proposed program for training Mexican biologists in banding will advance the collection of migration data (such as numbers, timing of migrations, body masses, plumage conditions, and route fidelity) and enhance the chances of recaptures so that we can learn more about flyway use. This would best be done in conjunction with an expanded banding program to gain significant knowledge of these pollinators' migration biology.

Wenink and Baker (1996) have succeeded in distinguishing breeding origins in composite flocks of migratory dunlins, *Caladris alpina* (=*Erolia alpina*), in five flyways also thought to reflect post-Pleistocene expansions. Hoping to bypass the slow wait for adequate banding recoveries, we are exploring use of DNA markers from feather papilla tissue to provide patterns that would permit us to subsequently distinguish the breeding origins of rufous and broad-tailed hummingbirds captured in migration and wintering populations, that is, distinguishing RUPNW from RURMt from breeding populations in Oregon, Washington, British Columbia, Alaska, and Montana and migration stops in Colorado, New Mexico, Arizona, and northern Mexico, and wintering populations in Mexico. This will make it possible to identify origins and learn about migration routes from a larger database than is possible from banding alone.

Using feather papilla samples from populations that we or collaborators have banded at times when seasonal residence is unequivocal—as confirmed by return of birds banded in past years—we have succeeded in extracting and purifying papillary DNA, which we submit to the polymerase chain reaction (PCR) using rufous mitochondrial primers (Calder, unpublished research). The Arizona Research Laboratories' Genomic Analysis and Technology Core facility provides both training and technical support in molecular biology, which my student interns and I have been utilizing to learn and perform the PCR work. This has led to the characterization of rufous mitochondrial loci Cytochrome b, Contol Regions I and II by DNA sequencing from very small, non-invasive sampling techniques. We have also developed several microsatellite DNA markers to detect the genetic variation in the autosomal chromosomes of both the rufous and the broad-tailed hummingbirds.

The following questions should be addressed:

- Do DNA sequences of RUPNW versus RURMt and BTUSA versus BTMEX differ intraspecifically in ways that have potential use for field studies of migration?
- What are the northward migration routes within Mexico?
- Which rufous (RUPNW or RURMt) fly south along the Pacific flyway, and what routes do they take within Mexico?
- What is the temporal sequence of migrants (RUPNW, RURMt, and sex-age distinctions) at points midway between breeding and wintering sites?
- What techniques and banding programs can yield the needed information? (a) Expansion of banding programs in northern Mexico, including Mexican biologists and possibly native para-ecologists, trained and supervised by Mexican or Arizona-Sonora Desert Museum (ASDM) field technicians will be needed. (b) Development and funding of DNA techniques will be needed for distinguishing origins of migrants, which now seems to require microsatellite DNA technology or the use of amplified fragment length polymorphism analysis to obtain sufficient discriminating genetic information.
- How much does the timing of migration vary from year to year, and what are the causes of the variation?
- How can the accuracy of monitoring be affected by year-to-year variation in migration onset, passage at a specific station, and arrivals?
- How can accuracy be improved in response to this seasonal variation?
- What insight can be gained into current problems and for motivating public concern if we have background appreciation of how migratory patterns evolve?
- What environmental conditions should be monitored to under-

stand nectar availability and migrant abundance? (a) Weather, soil moisture, and nectar standing crop and sugar concentration? (b) Small insect availability for electrolytes, fats, and amino acids not in nectar but necessary for pollinator maintenance?

Conservation

Of the 328 hummingbird species, 25 are considered to be threatened, and 1 is already extinct (Schuchmann 1999). Temperate species are not regarded as threatened because they tend to flourish early in "weedy" stages of succession. However, the rarity of some species with distributions barely extending northward into the United States may cause them to be classed as endangered by state agencies along the border. Watch-lists based on breeding bird surveys along routes affected by plant succession and perceptions of loss when birds are dispersed from feeders to abundant flowers may give false impressions of scarcity, as discussed above.

It is, however, important to recognize that birds are, in broad interspecies generalization, only one-sixth as populous as mammalian populations on an equal-body-size basis (Calder 2000). Stochastic factors, such as weather-related crises affecting floral resources, as noted above, can cause sudden drops in local populations. If more widespread (as would be possible during climate change), this could be disastrous, because if one metapopulation goes, so might go the neighboring meta-populations that could have supplemented the low reproductive output, as has been the case for broad-tailed hummingbirds. Hence the end could come quickly, without long-term trends to serve as warnings. When the pollinators crash, so may plant reproduction and its watershed-protecting nature.

ACKNOWLEDGMENTS

I am grateful for funding from the Arizona-Sonora Desert Museum through a grant from the Turner Foundation. Progress in identifying corridors in the spring of 2000 was possible due to the excellent planning and coordination of Tom Van Devender and Ana Lilia Reina, the floral and phonological expertise of Van Devender, Reina, Francisco Molina Freaner (UNAM-Hermosillo), and Mark Dimmit (ASDM), and the enthusiastic participation and keen observational skills of those persons and Karen Krebbs (ASDM), Susan Wethington (Univ. of Arizona), Lee Rogers, Reyna Castillo (Ud. de Sonora), Joan Day Martin (BBL Master Bander, NM Hummingbird Study Group), and Lucy Daley (ASDM docent).

Information on the wintering phase and pre-northward-migratory biology of the annual cycle has come from a very rewarding partnership with biologists of the Instituto Manantlán de Ecología y Conservacíon de la Biodiversidad of the Universidad de Guadalajara–Centro Universitario de la Costa Sur. For their insight, dedicated efforts, and friendship, I am indebted to Eduardo Santana C., Sarahy Contreras, Irma Ruan Tejera, Jorge Schöndube, José Carrillo, Rodrigo Esparza, Siux E. Díaz, Carla V. Blanco, Sandra Gallo, and Maria M. Ramírez.

Studies during the northbound migration were greatly facilitated by the staff of la Biosfera el Pinacate y Gran Desierto del Altar, the Reserva Biosfera Sierra de Pinacate, Biol. Carlos Castillo Sánchez, Director, and Guillermo Lara, Isabelle Granillo, and Victor Manuel Quiroga Soberanes.

LITERATURE CITED

Bailey, A. M., and R. J. Niedrach. 1965. Birds of Colorado, vol. 2, pp. 469–72. Denver Museum of Natural History, Colorado.

Buchmann, S. L., and G. P. Nabhan. 1996. The Forgotten Pollinators. Island Press, Washington, D.C.

Búrquez, A., A. Martínez-Yrízar, M. Miller, K. Rojas, M. Quintana, and D. Yetman. 1996. Mexican grasslands and the changing aridlands of Mexico: An overview and a case study in northwestern Mexico. In B. Yellamn, D. M. Finch, C. Edminster, and R. Hamre, eds. The Future of Arid Grasslands: Identifying Issues, Seeking Solutions, pp. 21–32. Proceedings RMRS-P-3, U.S. Forest Service, Rocky Mountain Station, Ft. Collins, CO.

Calder, W. A. 1985. Migration and population dynamics of hummingbirds. National Geographic Society Res. Rpts. 21:53–57.

———. 1987. Southbound through Colorado: Migration of rufous hummingbirds. National Geographic Research 3:40–51.

———. 1990. Avian longevity and aging. In Harrison, D. E., ed., Genetic Effects on Aging II, pp. 185–204. Telford Press, Caldwell, N.J.

———. 1993. The rufous hummingbird. In A. Poole and P. Stettenheim, eds., Birds of North America (no. 53). American Ornithologists' Union, Washington, D.C.; Academy of Natural Sciences, Philadelphia.

———. 1994. When do hummingbirds use torpor in nature? Physiological Zoology 67:1051–76.

———. 1999. Hummingbird migrations in Rocky Mountain meadows. In K. P. Able, ed.,

A Gathering of Angels: Migrating Birds and Their Ecology. Cornell University Press, Ithaca, N.Y.
———. 2000. Diversity and convergence: Scaling for conservation. In J. H. Brown and G. B. West, eds., Scaling in Biology, pp. 297–323. Oxford University Press, New York.
Calder, W. A., and L. L. Calder. 1992. The broad-tailed hummingbird. In A. Poole and P. Stettenheim, eds:, Birds of North America (no. 16). American Ornithologists' Union and Academy of Natural Sciences, Philadelphia.
———. 1994. The calliope hummingbird. In A. Poole and P. Stettenheim, eds., Birds of North America (no. 135). American Ornithologists' Union, Washington, D.C., and Academy of Natural Sciences, Philadelphia.
———. 1995. Size and abundance: Breeding population density of the calliope hummingbird. Auk 112:517–21.
Calder, W. A., and E. G. Jones. 1989. Implications of recapture data for migration of the rufous hummingbird (Selasphorus rufus) in the Rocky Mountains. Auk 106:488–89.
Calder, W. A., N. M. Waser, S. M. Hiebert, D. W. Inouye, and S. Miller. 1983. Site fidelity, longevity, and population dynamics of broad-tailed hummingbirds: A ten-year study. Oecologia 56:359–64.
Cox, G. W. 1985. The evolution of avian migration systems between temperate and tropical regions of the New World. American Naturalist 126:451–74.
Des Granges, J.-L. 1979. Organization of a tropical nectar feeding bird guild in a variable tropical environment. Living Bird 17:199–236.
Emerson, R. W. 1860. Beauty. In The conduct of life. Reprinted in The Collected Works of Ralph Waldo Emerson. Greystone Press, New York.
Finlay, J. C. 1999. Where were the rufous hummingbirds? Victoria Naturalist 55, 4:8–10.
Grant, K. A., and V. Grant. 1968. Hummingbirds and Their Flowers. Columbia University Press, New York.
Gryj, E., C. Martínez del Rio, and I. Baker. 1990. Avian pollen and nectar use in Combretum fruticosum (Loefl.). Biotropica 22:266–71.
Hill, G. E., R. S. Sargent, and M. B. Sargent. 1998. Recent change in the winter distribution of rufous hummingbirds. Auk 115:240–45.
Hixon, M. A., F. L. Carpenter, and D. C. Paton. 1983. Territory area, flower density, and time-budgeting in hummingbirds: An experimental and theoretical analysis. American Naturalist 122:366–91.
Immelmann, K. 1971. Ecological aspects of periodic reproduction. In D. S. Farner and J. R. King, eds., Avian Biology, vol. 1, pp. 341–89. Academic Press, New York.
Johnsgard, P. A. 1997. The Hummingbirds of North America. Smithsonian Institution Press, Washington, D.C.
Kearns, C. A., D. W. Inouye, and N. M. Waser. 1998. Endangered mutualisms: The conservation of plant-pollinator interactions. Annual Reviews of Ecology and Systematics 29:3–112.
Lange, R. S., and P. E. Scott. 1999. Hummingbird and bee pollination of Penstemon pseudospectabilis. Journal of the Torrey Botanical Society 126: 99–106.
Mayr, E. 1964. Inferences concerning the Tertiary American bird faunas. Proceedings of the National Academy of Science 51:280–88.

Mayr, E., and L. Short. 1970. Species taxa of North American birds. Publications of the Nuttall Ornithological Club 9:1–127.

Miller, A. H. 1963. Desert adaptations in birds. Proceedings, XIII International Ornithological Congress, pp. 666–674.

Muehter, V. R., ed. 1999. WatchList Website. Available online http://www.audubon.org/bird/watch/. National Audubon Society, New York.

Nabhan, G. P., and S. Buchmann. 1997. Services provided by pollinators. In G. C. Daily, ed., Nature's Services: Societal Dependence on Natural Ecosystems, pp. 133–150. Island Press, Washington, D.C.

Peterjohn, B. G., J. R. Sauer, and C. S. Robbins. 1995. Population trends from the North American breeding bird survey. In T. E. Martin and D. M. Finch, eds., Ecology and Management of Neotropical Migratory Birds: A Synthesis and Review of Critical Issues, pp. 3–39. Oxford University Press, New York.

Phillips, A. R. 1975. The migrations of Allen's and other hummingbirds. Condor 77:196–205.

Rathcke, B. J. 2000. Hurricane causes resource and pollination limitation of fruit set in a bird-pollinated shrub. Ecology 81:1951–58.

Russell, R. W., F. L. Carpenter, M. A. Hixon, and D. C. Paton. 1994. The impact of variation in stopover habitat quality on migrant rufous hummingbirds. Conservation Biology 8:483–90.

Schuchmann, K.-L. 1999. Family Trochilidae. In J. Del Hoyo, A. Elliott, and J. Sargatal, eds., Handbook of the Birds of the World, vol. 5: Barn-owls to Hummingbirds. Lynx Edicions, Barcelona.

Sibley, C. G., and B. L. Monroe. 1990. Distribution and Taxonomy of Birds of the World. Yale University Press, New Haven, Conn.

Tucker, V. A. 1974. Energetics of natural avian flight. In R. A. Painter, Jr., ed., Avian energetics. Publications of the Nuttall Ornithological Club 15, Cambridge, Mass.

Van Devender, T. R., K. Krebbs, A. L. Reina-Guerrero, M. Stephen, S. M. Russell, R. Russell, and W. A. Calder. 2000. Hummingbird plants in east-central Sonora, Mexico. Memorias II Simposium Internacional Sobre la Utilizacíon y Aprovechamiento de la Flora Silvestre de Zonas Aridas, pp. 203–8.

Wenink, P. W., and A. J. Baker. 1996. Mitochondrial DNA lineages in composite flocks of migratory and wintering dunlins (Caladris alpins). Auk 113:744–56.

CHAPTER 5

Migratory Patterns of the Rufous Hummingbird in Western Mexico

JORGE E. SCHONDUBE, SARAHY CONTRERAS-MARTÍNEZ, IRMA RUAN-TEJEDA, WILLIAM A. CALDER, AND EDUARDO SANTANA C.

Mexico represents a critically important wintering area for more than half of all migratory bird species that breed in the Nearctic (McNeely et al. 1990). Western Mexico is an ecologically and geographically complex area that harbors the highest diversity and abundance of migratory birds in the Neotropics during the winter (Hutto 1986; Palomera et al. 1994; Stotz et al. 1996). Of the 560 species of landbirds that can be found in this area, 34 percent are migratory species that arrive during the fall and winter to escape the cold winters at northern latitudes (Palomera et al. 1994).

Hummingbirds, warblers, and orioles come to western Mexico to escape the winter and to take advantage of the ecological bonanza represented by the millions of flowers that cover the mountains of the Sierra Madre Occidental from September to February. Species of sage, paintbrush, currant, morning glory, and composites cover meadows, abandoned agricultural areas, scrublands, and high-altitude grasslands, turning the mountain habitats into perfect wintering grounds for nectar-feeding birds (Ornelas and Arizmendi 1995; Arizmende 2001; Schondube et al., in press).

The high floral abundance in the mountains of western Mexico during the fall and winter months allows resident species of nectar-eating birds to breed and molt during that part of the year (Schondube et al., in press). Hummingbirds and other migratory birds that feed on floral nectar take advantage of this bounty to establish and maintain territories during the months that they spend in this part of the tropics. They use the energy contained in floral nectar to meet their metabolic needs, grow new feathers, build fat reserves, and refuel for their migration back north to their breeding grounds.

While visiting flowers, migratory nectar-feeding birds may act as pollinators. Their relative ecological importance as pollinators remains to be determined, but, because they actively move pollen from plant to plant while feeding and patrolling their winter territories, it has been assumed that they play an important role in the reproduction of several plant species.

One of western Mexico's most amazing migratory pollinators is the rufous hummingbird *(Selasphorus rufus)*. This species reaches the northernmost breeding latitude of any hummingbird and then migrates south in what is the longest known avian migration in relation to body size (Calder 1993). After breeding in the Pacific Northwest of the United States, Canada, and southern Alaska from April to July, individuals of this species migrate to central and western Mexico, where they spend the fall and part of the winter (Phillips 1975; Calder 1987; Calder and Jones 1989; Calder 1993). Rufous hummingbirds become abundant in western Mexico during November and December before starting to move back north in January.

What is the role of the rufous hummingbird as a pollinator of tropical and subtropical plant species on its wintering grounds? Is this species threatened by habitat destruction? Sadly, we do not have adequate answers to these questions. Little is known about the ecological role of this species once it flies south into Mexico. In this chapter we present preliminary information that we have gathered in western Mexico for the rufous hummingbird; we discuss its migratory patterns and speculate on the role of this species as a migratory pollinator.

We conducted our research in three mountain locations in the state of Jalisco, Mexico. Two of our study sites were located inside the Sierra de Manantlán Biosphere Reserve (Las Joyas Research Station, 1,900 m elevation, and El Almeal, 2,100 m elevation); the third site was in the Nevado de Colima National Park (3,100 m elevation). We began capturing and banding rufous hummingbirds in 1989 and continued to do so erratically through 1991. In the winter of 1991 we established two long-term mist-netting sites, adding another two in the summer of 1995. Our study sites were distributed in four habitat types: early successional secondary vegetation and old successional secondary vegetation where reforestation with pine trees occurred, both at Las Joyas; a high-altitude meadow with alders on the Nevado de Colima; and a meadow traditionally used for cattle-grazing (El Almeal). Because our three localities demonstrated the same patterns, we have pooled our data from them in the following interpretation.

Migratory Timing and Geographical Patterns

Rufous hummingbirds arrive in western Mexico as early as the end of September, and some individuals stay in the area as late as April. Dates of arrival for all years were between October 13 and 18. Birds arrive in small numbers in October and increase in numbers by the end of November through the beginning of December. Rufous hummingbirds are most abundant in December and January. In the context of Phillips's (1975) race-track pattern, this suggests an influx of north-bound transient rufous hummingbirds that spend their early winter farther south in the states of Michoacán, Guerrero, and Oaxaca, or perhaps to the east, in central Mexico. Confirmation of this scenario might emerge with the establishment of banding programs in those areas. We hypothesize that this increase in numbers of captures of rufous hummingbirds in Jalisco comes after eastern or southern populations initiate their migration northward, leaving earlier than those wintering in Jalisco or Colima (Phillips 1975; Calder 1993).

Males and females arrived at our study areas at the same time. Because adult males leave the breeding grounds before females (Calder 1987; Colwell 1992), our results suggest that females catch up with males sometime along the southbound migration. The proportion of females increased in January and February without any significant increase in total rufous capture rates, suggesting that males depart from the area earlier than females. Rufous hummingbirds captured in March tended to be females or juvenile males, and all the birds captured in April were females, supporting the hypothesis that adult males migrate northward first (Calder 1987; Colwell 1992; Calder 1993).

Female rufous hummingbirds outnumbered males in our net captures during almost all of the months we sampled. Sex ratios were 2.5 females per male (± 2 SD, n = 48 months), and during five months we only captured females. There are two possible explanations for this: an altitudinal gender-segregation in the rufous hummingbird in our study area, with adult males establishing territories at lower altitudes than adult females or juveniles; or behavioral differences between males and females. Females could be easier

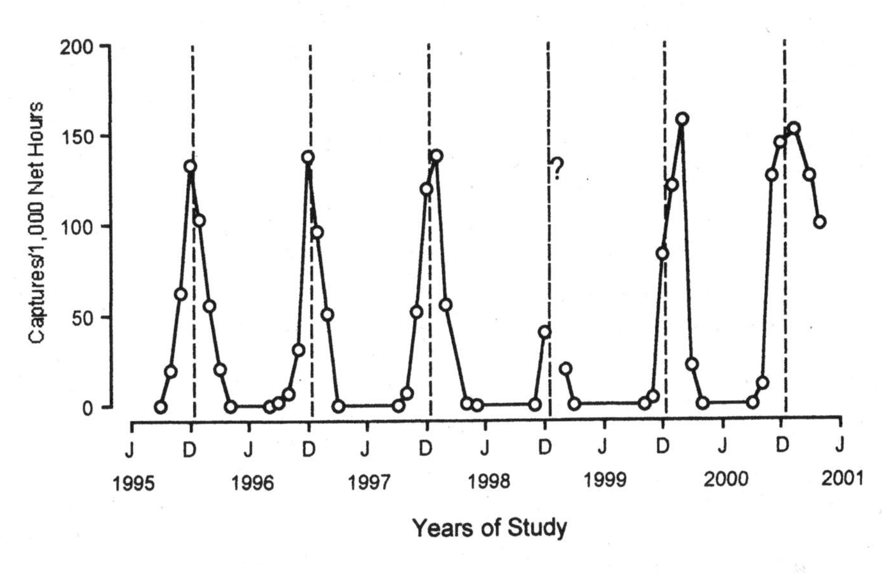

Figure 5.1. Relative abundance of rufous hummingbirds in western Mexico from mist-netting data. The birds arrive in low numbers during the month of September. Numbers increase rapidly during November and December, and tend to peak in January, when transient hummingbirds start to migrate through the area. Abundance was similar for six years of data, suggesting stable population sizes. J = June; D = December; broken vertical lines separate years.

to capture than males if they were nomadic. If adult males hold onto high-quality territories, pushing females and juveniles to suboptimal ones (Gass 1978), then both females and juveniles would tend to move more while foraging, increasing their vulnerability to capture.

Juvenile males tended to arrive earlier at our mountain sites than adult males, whereas there was no age-bias in arrivals of adult and juvenile females. Because juveniles are losing their juvenile plumage and bill striations in January and February, the adult versus juvenile distinction becomes progressively more difficult. This makes it impossible to determine whether adults with previous residency in the area departed earlier than first-year birds.

Numbers of rufous hummingbirds drop rapidly during February when presumed transient individuals have passed on and locally wintering individuals start to migrate northward (figure 5.1). Decreased captures and sightings suggest that departure dates for birds using western Mexico as wintering grounds were mostly between February 20 and March 21. Rufous hummingbird abundance in March is low, and only a few stragglers keep territories to the end of the month, though we have observed some even staying until April.

Migratory Routes and Migration Strategies within Mexico during the Winter

Migration strategies, wanderings, and routes of the rufous hummingbird within its wintering range are even more poorly known than are the routes north of the Mexico-U.S. border. Phillips's (1975) analysis of collection dates and locations from skin specimens in ornithological collections suggested an elliptical southeast-western-northward annual routing that is consistent with weather patterns and floral phenology (figure 5.2). Over the Mexican Central Plateau and the Sierra Madre Oriental, they may wander as they move west, passing through Jalisco and Michoacán in January before migrating northward along the Pacific Coast and the Sierra Madre Occidental.

However, this is not the only migratory route used by rufous hummingbirds in North America. Some populations migrate south along the Pacific coastal ranges and the Sierra Nevada in the United States (Gass 1978; Carpenter and Hixon 1988). The percentage of rufous that migrates south using this western route is unknown, but it seems that higher numbers of birds use the Rocky Mountain–central Mexico route to migrate south (Calder 1987). Recapture data from the Rocky Mountain route suggests that individuals have a high fidelity to this route (Calder 1987; Calder and Jones 1989); even though we lack recapture data for the Pacific route, a similar scenario can be expected.

Dates of arrival of rufous hummingbirds at our study sites in western Mexico indicate the use of a migration route other than Phillips's ellipse. Because rufous hummingbirds arrive in September and October more or less simultaneously in both central and western Mexico, it seems that southbound birds use a migration pattern that looks more like a fan than an ellipse (see figure 5.2). As rufous hummingbirds migrate south following a central route, they may reach a point in the Mexican Central Plateau where they disperse either south, to central Mexico, or west, toward the Pacific coast.

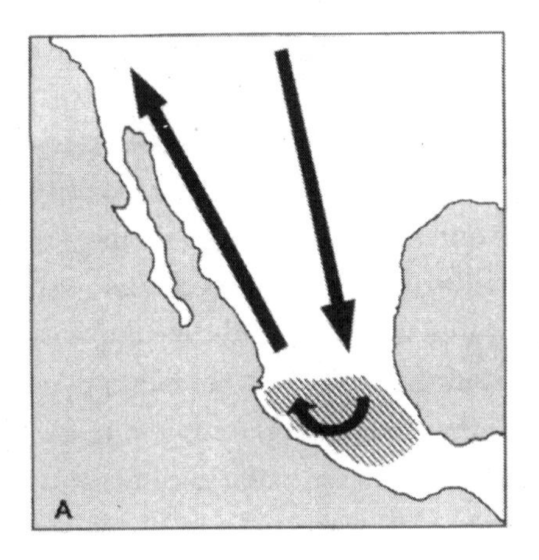

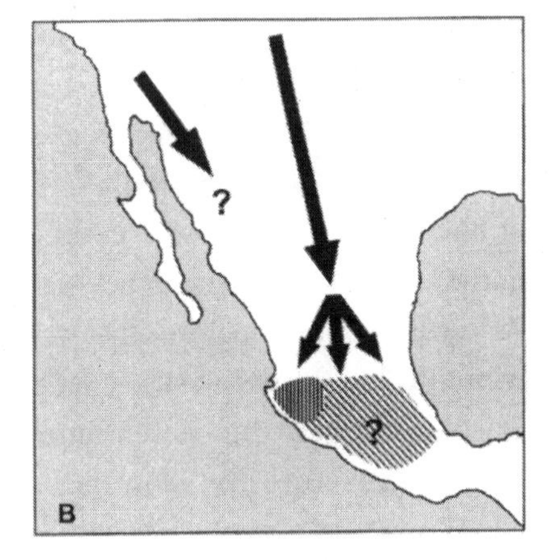

Figure 5.2. Migratory routes of the rufous hummingbird in Mexico. (A) Phillips's (1975) hypothesis of the rufous hummingbird migration. Southbound migrating birds use the central Mexico plateau and the eastern Sierra Madre to reach central Mexico in September. Birds wander in their wintering grounds (diagonal lines) and move west before starting the northbound migration in January and February. (B) Our data suggest that birds migrate south in a fan-like fashion, reaching western and eastern central Mexico simultaneously. In western Mexico, they establish residency territories that last all winter (vertical lines). It is unknown if this is the case for other areas of their wintering grounds. There is no information about the Pacific southbound migration route.

Unfortunately, our coverage is too limited to use arrival dates and recapture data to evaluate the importance of the southbound route between the Pacific coast and the Sierra Madre Occidental. To do that, monitoring stations in the states of Nayarit, Sinaloa, and Sonora, would be needed.

The Mountains of Jalisco as Wintering Grounds and Stopover Sites

Phillips (1975) proposed that rufous hummingbirds wandered during the winter, with birds following flowering phenology first south into central Mexico and then westward. Recapture data from our three study sites suggest that in western Mexico the south-bound migration ends with the establishment of residence areas, rather than with a continual wandering behavior through the winter. It seems plausible that rufous hummingbirds arriving in September and October in central Mexico would also exhibit winter territorial behavior.

Rufous hummingbirds exhibited site fidelity and maintained territories over periods of several months at our three study sites. The majority of birds that were captured in different wintering seasons were recaptured in the same location where they were initially captured (89 percent, $n = 97$). Also, rufous hummingbirds that were captured several times during the same wintering season were usually recaptured in the same mist-net location, or adjacent to where they were captured the first time (78 percent, $n = 348$).

Recapture data show that winter territories are maintained for periods of up to five months—from September to February, or from November to March or April. Both males and females maintained winter territories, regardless of their age. They established territories and defended them from both resident and migratory nectar-eating birds the entire time they stayed in the area. Des Granges (1979) described rufous hummingbirds as "subdominant" relative to the local species of hummingbirds and nectar-eating passerines. Our observations and recapture data show that, even if rufous hummingbirds allow some large species of nectarivorous birds to temporally displace them from foraging areas, they hold territories and defend them aggressively against other rufous and small resident hummingbirds like the bumblebee, the white-eared, and the green-violet-eared.

It is not clear whether all the rufous hummingbirds that winter in western Mexico establish territories. The high numbers of birds that we captured only once suggest two other complementary patterns of wintering behavior: local wandering behavior, and/or the use of the area as a stopover site, both during the south- and north-bound migrations. Birds that were unable to secure a territory for the wintering season could engage in nomadic behavior that makes them unlikely to be recaptured. They could be "sneaking" into other hummingbirds' territories to feed and then moving to other areas. Another option is that birds establishing territories in low-quality habitats with poor flower production must become nomadic and look for new territories when the flower supply dwindles.

For each subsequent year of study, we had an increase in the number of captured rufous hummingbirds during January. Phillips suggested that January is the time when hummingbirds that spend their winter in central

Mexico move west before migrating northward. The majority of our January birds were captured only once. This suggests that they were not using the area for long periods of time. If these birds represent individuals migrating northward, they could be using the area as an stopover site to refuel. The fact that most birds captured only once during January had lower fat loads than "resident" wintering individuals supports this idea, and it indicates that the mountains of southern Jalisco could be an important area for refueling while en route to the northern breeding grounds.

Body Weight and Fat Accumulation Patterns

Body weights of rufous hummingbirds varied during the time they spent in western Mexico in relation to three events: arrival at wintering grounds after south-bound migration; residency on wintering grounds; and "preparation" to migrate northward. Newly arrived birds during October and November exhibited lower body masses (3 g ± 0.51, n = 181) than the birds captured during the rest of the winter months (3.4 ± 0.51, n = 1,291). After birds established territories, they tended to gain some mass (0.3–0.8 g) and then maintain weight until the end of January. During their residency on the wintering grounds, rufous hummingbirds undergo a complete molt in which they replace all of their body, wing, and tail feathers (Calder 1993; Ruan-Tejeda, unpublished data). The fact that the birds did not gain weight during the months of December and January could be because of the extra energetic expense caused by the molt. In February, after the molting is completed, they start to gain weight again in preparation for their north-bound migration.

These changes in body mass were correlated with accumulation of subcutaneous fat. Birds of both sexes arrived with small loads of fat in their furcular and abdominal regions (only between 0 and 15 percent of the furcula covered with fat). The amount of subcutaneous fat exhibited by rufous hummingbirds increased progressively during November, became constant during December and most of January, and then increased again at the end of January and February (up to 75–100 percent of the furcula covered with fat by February). Males arrived with lower fat reserves than females (0–25 percent versus 25–50 percent of the furcula covered with fat), but both sexes left the area with similar fat loads. Juveniles showed lower fat reserves than adults during their stay in the area, which suggests that they could have been using lower quality territories for feeding.

Body masses at capture show a day-long gain trend, which becomes steeper as the season advances (figure 5.3). Rufous hummingbirds start feeding just before sunrise and continue to actively feed during the entire day. As the day passes they gain weight by slowly accumulating subcutaneous fat (Calder and Contreras-Martínez 1995). Fat accumulation averages 0.011 g/h (± 0.005 SD, n = 2,212) for any given day of the winter, for a total gain of 0.12 g through the 11.3 h feeding day.

Patterns of daily changes in body mass differed during the winter months. Birds captured during the months of October, November, December, and February increased their body mass 0.015 g/h (± 0.04 SD, n = 923) during the day, whereas birds captured in January maintained constant body masses (3.2 g ± 0.22 SD, n = 776; see figure 5.3). During March, captured birds gained mass as the day progressed at a faster rate (0.042 g/h ± 0.01 SD, n = 123). Total fat accumulated per day varied from 0 g in January to around 0.16 g in October, November, December, and February, and 0.47 g in March. This could be due to changes in the photoperiod length, thermoregulatory demands, competition with both migratory and resident hummingbirds, or a combination of all these factors.

Regressions of body masses upon capture time shows the population's trend to store energy for the winter nights (see figure 5.3). With lengthening days and increasing food availability, the rate at which body mass increased became faster. The stored energy necessary to maintain high body temperatures (versus entering torpor for fat conservation) was not attained until late February, when the rufous hummingbirds gained the mean equivalent of 11.2 kJ/d, as calculated from ambient temperatures and laboratory data for metabolism as a function of temperature. During January birds accumulate little fat, suggesting the use of hypothermic torpor to survive the long cold nights (11 h). During the first two weeks of February, fat accumulation was enough to maintain body temperature for almost half the night (5.5 h). At the end of the month, birds were able to store enough fat to spend the whole night without entering torpor. Rufous hummingbirds could be using torpor regularly in order to save and accumulate fat for migration (Carpenter and Hixon 1988). Observations of torpid rufous hummingbirds during the

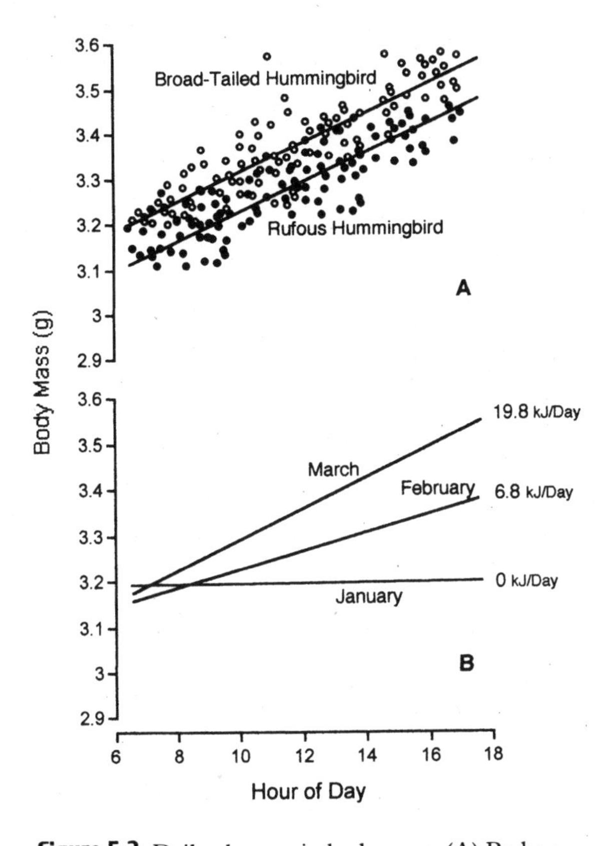

Figure 5.3. Daily changes in body mass. (A) Body mass gains in populations of rufous hummingbirds (filled circles) and broad-tailed hummingbirds (open circles) are shown as functions of time of capture for a banding site in Jalisco (Nevado de Colima, Dec. 22–26, 1992). The slopes of the regressions are statistically indistinguishable, indicating similar feeding success for these two species of the same size and genus. (B) Body mass gain rates (regression slopes for body masses as functions of capture times) of rufous hummingbirds increase significantly and progressively with early seasonal lengthening of photoperiod and apparent floral abundance in mountain sites of Jalisco. Only weight gains of > 11.2 kJ per day would have met our calculated energy cost of maintaining high (non-torpid) and continuous homeothermic body temperatures all night.

months of January and February in our study areas support this possibility. By March the mean temperature rises, helping the birds that remain in the area gain weight rapidly in preparation for migration.

Plant Species Visited by Rufous Hummingbirds in the Mountains of Western Mexico

The relative importance of rufous hummingbirds as pollinators on their wintering grounds remains unknown. Our knowledge of their diet is limited, and no studies have been conducted to determine their efficiency as pollinators on their wintering grounds. The documented diet of this species in the mountains of western Mexico includes the nectar of forty-one plant species in thirteen families (table 5.1). The families Lamiaceae and Asteraceae are the most commonly represented among the flowering species visited by rufous hummingbirds (eleven and six species, respectively). Curiously, only seventeen of all the plant species visited by the rufous hummingbird exhibited flowers with characteristics generally assumed to indicate specialization for hummingbird pollination, such as tubular corollas and reddish colors (Grant and Grant 1967, 1968; Stiles 1976). The other twenty-four species have open cup flowers with white, yellow, blue, or purple colors.

This wide diversity of floral shapes and colors exhibited by flowers that rufous hummingbirds visited suggest that this species may be acting as a foraging generalist in its wintering grounds. When confronted with flowers with corollas too long for them to extract the nectar though the corolla opening, rufous hummingbirds can become nectar "robbers" (*sensu* Inouye 1980). For example, in order to extract nectar from the long corollas of *Ipomea orizabensis*, they use the tips of their beaks as scissors to cut the base of the corolla and drink the nectar through the opening. The ability to extract nectar from flowers with different forms of corollas, sizes, and colors could be an important adaptation for a migratory species facing contrasting habitats en route.

Unanswered Questions

The migratory and wintering ecology of the rufous hummingbird south of the U.S.–Mexico border remains largely unknown. We have learned that, in western Mexico, some individuals establish territories that are used

TABLE 5.1. Plant Species Visited by the Rufous Hummingbird in the Mountains of Western Mexico (1,000–3,400 m elevation)

Family	Species
Asteraceae	*Cirsium mexicanum*
	Cirsium nivale
	Cirsium tolucanum
	Senecio angustifolius
	Senecio barba-johannis
	Senecio reticulatus
Buddlejaceae	*Buddleja cordata*
Convolvulaceae	*Ipomoea arborescens* (lower elevations)
	Ipomoea hederifolia
	Ipomoea orizabensis (robbed)
Ericacea	*Arbutus occidentalis*
	Arbutus xalapensis
Labiatae (Lamiaceae)	*Lepechinia nelsonii*
	Lobelia laxiflora
	Salvia cinnabarina
	Salvia elegans
	Salvia gesneraeflora
	Salvia iodantha
	Salvia lavanduloides
	Salvia longispicata
	Salvia mexicana
	Stachys bigelovii
	Stachys coccinea
Leguminosae	*Calliandra formosa*
	Calliandra grandiflora
Liliaceae	*Agave pedunculifera*
Onagraceae	*Fuchsia encliandra*
	Fuchsia fulgens
	Fuchsia microphylla
Rosaceae	*Prunus persica* (introduced species)
	Prunus serotina
	Rubus adenotrichos
	Rubus glaucus
Rubiaceae	*Crucea coccinea*
Saxifragaceae	*Ribes ciliatum*
Scrophulariaceae	*Castilleja scorzoneifololia*
	Castilleja spp.
	Penstemon roseus
	Penstemon spp.
Solanaceae	*Cestrum terminale*
	Nicotiana glauca (introduced species, lower elevations)

and defended all winter, and that they are able to "hold their own" in the highly diverse communities of resident hummingbirds. We are learning which plants they visit in their wintering habitats and stopover sites, and assessing their abilities to forage on a large diversity of plants with contrasting floral morphologies (see chapter 6 in this volume). The rufous hummingbird is a migratory species "belonging" to three countries (Mexico, Canada, and the United States) and is thought by some to be suffering population declines in parts of its range. Knowledge of its ecology could have important conservation consequences.

A prerequisite for protecting the rufous hummingbird is unveiling the mysteries of its migratory routes. We have focused on studying this species on its breeding grounds and have some information about its wintering ecology, but we still know little about what happens in between the two extremes of its distribution. Where do the rufous hummingbirds that use the Pacific coast or the Mexican Central Plateau routes to migrate southward originate? Do birds in central Mexico, Michoacán, Guerrero, and Oaxaca establish winter territories? It will be necessary to sample along the core sections of the route and to record dates of passage and relative abundance of the birds before we can fully understand the migratory patterns of the species.

We also need to understand the importance of stopover sites and the ability that this species has to find them. Recaptures of banded individuals en route in consecutive years suggest both route and stopover site fidelity (Calder and Jones 1989; Calder 1993). Several researchers have reported that it may take one to two weeks at a stopover for rufous hummingbirds to rebuild burned fat reserves and recover muscle mass after a long flight (Carpenter et al. 1983, 1993; Russell et al. 1994). Counting the time spent in these stops, a one-way migration from breeding grounds in Canada to central Mexico will be close to ten weeks long (Calder 1987). We need to learn how sensitive rufous hummingbirds are to human alteration of their traditional stopover sites. If rufous hummingbirds need to use specific stopover sites, we will need to locate and protect them to ensure the continuity of the migration.

A second set of conservation-related questions has to do with the changes in land use along the areas that we know rufous hummingbirds use for migration and as wintering grounds. Does the conversion of natural vegetation to agricultural production, artificial grasslands, or human settlements pose a serious threat to rufous hummingbirds? In the mountains of western Mexico, the main causes of habitat destruction are cattle ranching, logging, forest fires, and slash-and-burn agriculture. These activities tend to generate open areas covered by secondary vegetation that is frequently used by rufous hummingbirds. It may be that habitat management under moderate disturbance regimens in mountain habitats can be beneficial to this species. Transformation of habitats along riparian corridors, mountain and lowland valleys, and hillsides could have a relatively more dramatic effect on rufous hummingbirds. The conversion of natural vegetation to agriculture could destroy important habitats used during migration, reducing the availability of stopover sites, particularly in Sinaloa and Nayarit. The conversion of deciduous tropical forest to pasture and agricultural fields along the Pacific coast of Mexico could result in negative consequences for the north-bound migration, or for the populations that migrate southward using the west coast flyway.

Finally, we need to understand the ecological importance of this species on its wintering grounds. The other side of the coin is that hummingbirds may be significant pollinators of plant species that are important in the re-vegetation of grazed, logged, and slashed-and-burned lands and, therefore, to watershed recovery and protection. Is the rufous hummingbird an effective pollinator during its stay in the wintering grounds? What is the relative importance of this species as a pollinator when compared with the resident species of hummingbirds, nectar-feeding passerines, and insects that visit the same flowers? What is the scale at which the rufous hummingbird moves pollen between conspecific plants? Do they move pollen within a patch, between patches, or among populations of plants? Answering these questions will allow us to understand and protect this incredible species, whose migration connects North America from Alaska to Oaxaca.

ACKNOWLEDGMENTS

These results are part of the Manantlán bird-monitoring program directed by Eduardo Santana C. The project was financed by grants awarded to him from the Universidad de Guadalajara, National Fish and Wildlife Foundation, Denver Audubon Society, Allen Stokes, U.S.-A.I.D., Grace J. Calder Trust, General Services Foundation, Colorado Wildlife Heritage

Foundation, and Paul and Bay Foundations. Long-term collaboration with Borja Mila of Point Reyes Bird Observatory was essential in developing field techniques. Jorge E. Schondube received field training at the U.S. Forest Service's Redwood Sciences Laboratory in California and financial support from a CONACyT doctoral scholarship. We thank Blanca Claudet Guerrero, Jose Guadalupe Carrillo, and Rodrigo Esparza for their intensive help in the field. They provided friendship and ideas in various parts of the study. We also want to thank the staff of Las Joyas Research Station: Deidad Partida, Francisco Hernández-Vázquez, Ruben Ramírez, José Aragón, and the cooking team (Ludivina Cruz, Meche Cruz, Irene Cruz, and Doña Ofelia Cruz) for logistic support and fantastic meals. Todd McWhorter read and criticized the manuscript. This chapter is dedicated to all of the volunteers who, for seven years, worked with us setting up mist-nets.

LITERATURE CITED

Arizmendi, M. d. C. 2001. Multiple ecological interactions: Nectar robbers and hummingbirds in a highland forest in Mexico. Canadian Journal of Zoology 79:997–1006.

Calder, W. A. 1987. Southbound through Colorado: Migration of rufous hummingbirds. National Geographic Research 3:40–51.

———. 1993. Rufous hummingbird (*Selasphorus rufus*). Pp. 1–18 in A. Poole and F. Gill, eds., The Birds of North America, No. 53. The Academy of Natural Sciences, Philadelphia; The American Ornithologists' Union, Washington, D.C.

Calder, W. A., and S. Contreras-Martínez. 1995. Migrant hummingbirds and warblers on Mexican wintering grounds, Pp. 123–38 in M. H. Wilson and S. A. Sader, eds., Conservation of Neotropical Migratory Birds in Mexico. Maine Agricultural and Forest Experiment Station.

Calder, W. A., III, and E. G. Jones. 1989. Implications of recapture data for migration of the rufous hummingbird (*Selasphorus rufus*) in the Rocky Mountains. Auk 106:488–89.

Carpenter, F. L., and M. A. Hixon. 1988. A new function for torpor: Fat conservation in a wild migrant hummingbird. The Condor 90:373–78.

Carpenter, F. L., M. A. Hixon, C. A. Beuchat, R. W. Russell, and D. C. Paton. 1993. Biphasic mass gain in migrant hummingbirds—Body-composition changes, torpor, and ecological significance. Ecology 74:1173–82.

Carpenter, F. L., D. C. Paton, and M. A. Hixon. 1983. Weight gain and adjustment of feeding territory size in migrant hummingbirds. Proceedings of the National Academy of Sciences of the USA 80:7259–63.

Colwell, R. 1992. Rufous hummingbirds at CCRS: A summary of four years of banding data. Riparia News 7:1–5.

Des Granges, J.-L. 1979. Organization of a tropical nectar feeding bird guild in a variable tropical environment. Living Bird 17:199–236.

Gass, C. L. 1978. Rufous hummingbird feeding territoriality in a suboptimal habitat. Canadian Journal of Zoology 56:1535–39.

———. 1979. Territory regulation, tenure, and migration in Rufous hummingbirds. Canadian Journal of Zoology 57:914–23.

Grant, K. A., and V. Grant. 1967. Effects of hummingbird migration on plant speciation in the California flora. Evolution 21:457–65.

———. 1968. Hummingbirds and Their Flowers. Columbia University Press, New York.

Hutto, R. L. 1986. Migratory landbirds in western Mexico: A vanishing habitat. Western Wildlands 11:12–16.

Inouye, D. W. 1980. The terminology of floral larceny. Ecology 61:1251–53.

McNeely, J. A., K. R. E. Miller, W. V. Reid, R. A. Mittermeier, and T. B. Werner. 1990. Conserving the World's Biological Diversity. IUCN, Gland, Switzerland; WRI, CI, WWF-US, and the World Bank, Washington, D.C.

Ornelas, J. F., and M. C. Arizmendi. 1995. Altitudinal migration: Implications for the conservation of the neotropical migrant avifauna of Western Mexico. Pp. 98–112 in M. H. Wilson and S. A. Sader, eds., Conservation of Neotropical Migratory Birds in Mexico. Maine Agricultural and Forest Experiment Station.

Palomera, C., E. Santana C., and R. Amaparan-Salido. 1994. Patrones de distribucion de la avifauna en tres estados del occidente de Mexico. Anales del Instituto de Biologia UNAM. Serie Zoología (65):137–75.

Phillips, A. R. 1975. The migrations of Allen's and other hummingbirds. Condor 77:196–205.

Russell, R. W., F. L. Carpenter, M. A. Hixon, and D. C. Paton. 1994. The impact of variation in stopover habitat quality on migrant rufous hummingbirds. Conservation Biology 8:483–90.

Schondube, J. E., E. Santana C., and I. Ruan-Tejeda. In press. Biannual cycles of the cinnamon-bellied flowerpiercer. Biotropica.

Stiles, F. G. 1976. Taste preferences, color preferences, and flower choice in hummingbirds. Condor 78:10–26.

Stotz, D. F., J. W. Fitzpatrick, T. A. Parker III, and D. K. Moskovits. 1996. Neotropical Birds: Ecology and Conservation. The University of Chicago Press, Chicago.

CHAPTER 6

Hummingbird Plants and Potential Nectar Corridors of the Rufous Hummingbird in Sonora, Mexico

THOMAS R. VAN DEVENDER, WILLIAM A. CALDER, KAREN KREBBS, ANA LILIA REINA G., STEPHEN M. RUSSELL, AND RUTH O. RUSSELL

Hummingbirds in the family Trochilidae play colorful and functionally important roles in pollination services of northwest Mexico; sixteen species have been reported for Sonora alone (Russell and Monson 1998), including both resident and migratory species. In Baja California, Chihuahua, and Sonora, rural people call them *chuparrosas* (often misspelled with one 'r'), whereas urban dwellers and national public school curricula call them *colibríes*, the French name (Johnsgard 1997). Elsewhere in Mexico and Latin America, they are called *chupaflor*, *chupamirto*, *picaflor*, and other names. In northwestern Mexico, individual species of hummingbirds typically do not have unique common names, although the rufous hummingbird *(Selasphorus rufus)* is called *chuparrosa amarilla* near the Cascada de Basaseachi in Chihuahua. Hummingbirds are seasonally important pollinators of numerous native and introduced plants, competing with bees, butterflies, hawkmoths, and passerine birds for their nectar. Although hummingbirds occur throughout Sonora, their diversity and abundance are greater in environments where extreme drought and heat occur less frequently. Because rainfall increases as temperature decreases with higher elevations, the driest, hottest areas in Sonora are along the Sea of Cortés and the wettest, coolest are in the higher mountains of the eastern Sierra Madre Occidental. There is also a steep north-south gradient of hummingbird habitat suitability, as the total and annual percentages of summer rainfall increase toward the New World tropics. This means that hummingbirds are more likely to be found in wetter areas in the more tropical areas of southeastern Sonora and, secondarily, in the highlands of northeastern Sonora (south of the Chiricahua, Huachuca, and

Santa Rita Mountains of Arizona). However, the numbers and species of hummingbirds present at any given site vary greatly between seasons as important nectar plants initiate or terminate their flowering, or as migratory species pass through. The timing of peak flowering of some of these plants is loosely correlated with the peak of migrating hummingbirds (Waser 1979). In this chapter, we discuss our observations of the plants in Sonora visited by hummingbirds with special emphasis on migratory rufous hummingbirds in Sonora during 1999–2001. Our objective is to hypothesize the probable locations of nectar corridors running through northwestern Mexico to the United States on the basis of floral distributions and hummingbird behaviors.

Methods

Based on previous studies, we selected observation sites in Sonora that could potentially help discern migratory corridors for hummingbirds on the basis of abundant populations of important nectar plants. Sites near Mexico Highway 15 from Guaymas to Hermosillo to Imuris, Sonora, were presumed to be along a spring migration route from tropical Sinaloa and southern Sonora, northwestward to California. We assume that rufous hummingbirds used this route to reach their early summer breeding grounds in the Pacific Northwest, Canada, and Alaska (Phillips 1975; Calder 1993). Sites along the Sea of Cortés visited were San Carlos, Bahía Kino, Punta Chueca, and Puerto Lobos. Sites along Mexico Highway 16 from Hermosillo southeast to Yécora formed a west-east transect designed to intercept northward or southward migrating hummingbirds that might utilize more easterly routes (figure 6.1). The transect followed the vegetational gradient from desertscrub in the Plains of Sonora subdivision of the Sonoran Desert near Hermosillo (300–600 m elevation) through foothills thornscrub near the Río Yaqui (180 to 300–400 m) and tropical deciduous forest near Tepoca (300–400 to 1,200 m) to oak woodland near La Palmita (1,200–1,600 m) and pine-oak forest on Mesa del Campanero (1,600–2,200 m) in the Sierra Madre Occidental (Búrquez M. et al. 1992; Reina G. et al. 1999). This transect was visited thirteen times between December 1999 and September 2001 by teams of three to nine people before, during, and after the spring (February–May) and during the summer (July–September) migrations of rufous hummingbirds (Phillips 1975; Calder 1993).

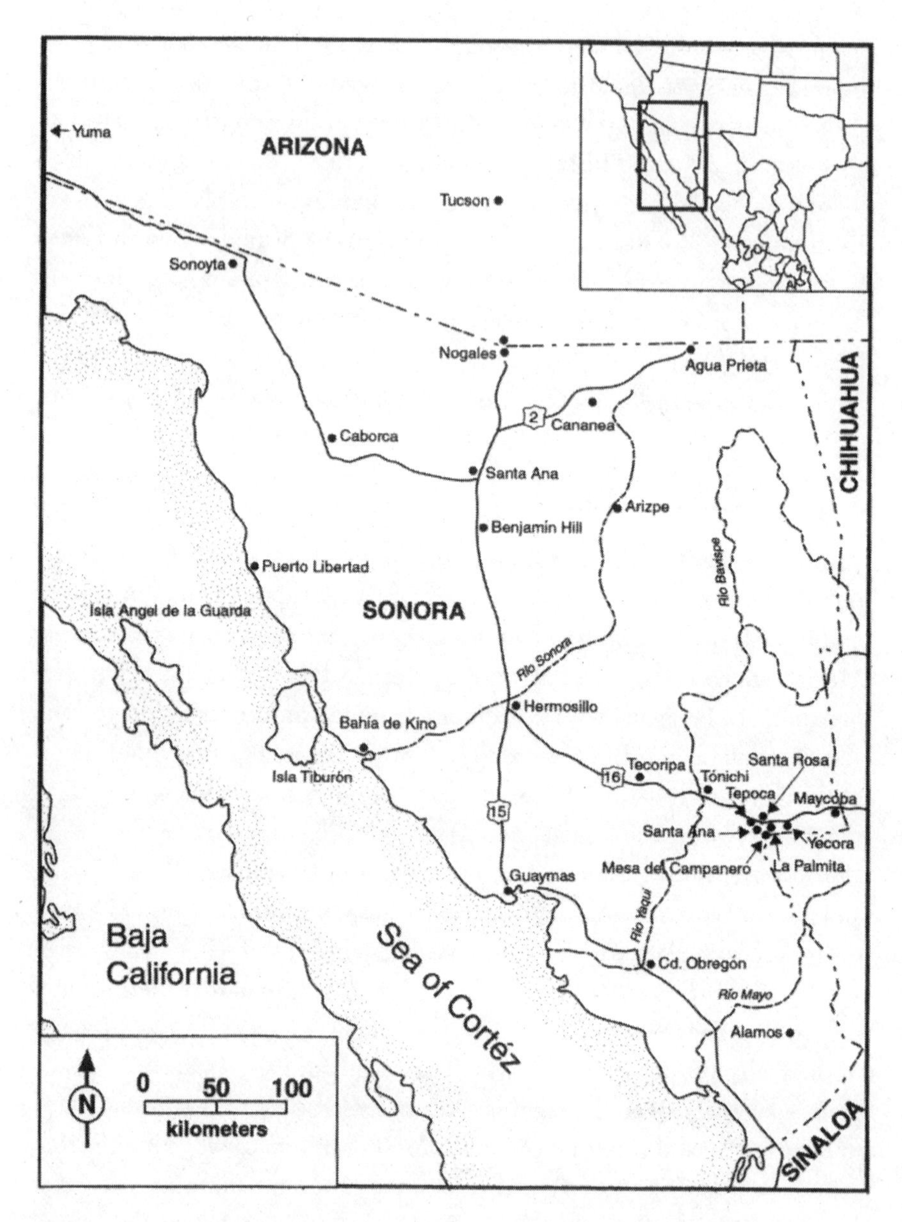

Figure 6.1. Geographic features and localities discussed in this chapter.

At each locality, we observed all species of hummingbirds present, and their behaviors, including floral visitation, were recorded for at least an hour. The behaviors we documented included visitation at flowers and feeders, insect foraging (at flowers, gleaning, hawking, and sallying), roosting in plants or on nests, aggressive interactions, and courtship flights. Our field notes, which include species, locality, and date, are maintained in a database at the Arizona-Sonora Desert Museum. Separate records were entered for each species of plant visited by a hummingbird species for each locality and observation day. In addition, we have included eleven hummingbird observations from the December 22, 1999, Yécora Christmas Bird Count (Jack Whetstone, personal communication, 2001). The Christmas Bird Count area surveyed was centered at Santa Rosa in the Municipio de Yécora with a 24 km radius that crossed Mexico 16, the highway to Chihuahua City. Also included were observations of hummingbirds by teachers from *telesecundarias* (satellite middle school) in Curea and Tónichi made during their Migratory Pollinators project and plant phenology studies and by Seri Indian Para-Ecólogos.

We recorded phenological condition of all perennial plants at each locality. We then deposited voucher specimens of species of plants visited by hummingbirds into the herbaria at the University of Arizona, the Universidad de Sonora, and the Universidad Nacional Autónoma de México (MEXU). Flowers were collected for a pollen reference collection in the Instituto de Ecología-UNAM in Hermosillo, currently being developed by Cristina Peñalba for use in identifying pollen samples removed from hummingbird feathers and beaks.

Results

We made 499 observations (OBS) of twelve species of hummingbirds on ten trips into Sonora. Individual observations ranged from one to forty-nine birds; thus, the observations represent the simple presence of a species and do not reflect relative abundances. Numbers of observations are a rough estimate of the frequency of occurrence. We observed hummingbirds visiting flowers of eighty-four species of plants including fifty-four native, twenty-eight cultivated, and two introduced wild species. The most commonly visited plants were tree ocotillo/ocotillo macho (*Fouquieria mac-*

dougalii, seventy-five OBS), tree tobacco/juanloco (*Nicotiana glauca*, forty-two OBS), pineapple sage (*Salvia elegans*, thirty-seven OBS), Texas betony (*Stachys coccinea*, thirty-seven OBS), tree morning glory/palo santo (*Ipomoea arborescens*, thirty OBS), wild jícama (*I. bracteata*, fourteen OBS), limita (*Anisacanthus andersonii*, eleven OBS), hierba del piojo (*Mandevilla foliosa*, ten OBS), rama del toro/chuparrosa (*Justicia candicans*, seven OBS), Madrean beardtongue (*Penstemon wislizenii*, six OBS), Maycoba sage (*Salvia betulaefolia*, six OBS), ocotillo (*Fouquieria splendens*, six OBS), and uvalama (*Vitex mollis*, six OBS). Of these twelve species of plants common in Sonora, the ranges of only Texas betony, rama del toro, ocotillo, and tree tobacco reach into Arizona, providing nectar for hummingbirds migrating northward from Mexico.

Sonoran Plants Visited by Hummingbirds

A total of nineteen species of plants were previously reported to have been visited by hummingbirds in Sonora (Gentry 1942; Marshall 1957; Calder 1993; Baltosser and Scott 1996; Johnsgard 1997; Russell and Monson 1998; Powers and Wethington 1999). Several of these species were reported as commonly visited by hummingbirds in Sonora: ocotillo and Mohave beardtongue *(Penstemon pseudospectabilis)* in northwestern Sonora, chuparrosa *(Justicia californica)* along the coast of the Sea of Cortés, tree morning glory in tropical deciduous forest, and hierba del piojo in oak woodland in the Madrean highlands of the east. The introduced tree tobacco is found in disturbed habitats up to 1,800 m elevation.

Of the seventy-seven species of plants that we observed hummingbirds visiting (see appendix 6.1), fifty-eight of these species have not been reported previously for Sonora. Nine additional species were recently reported to us by field biologists in Sonora: cardón *(Pachycereus pringlei)*, organpipe cactus *(Stenocereus thurberi)*, and sahuaro (*Carnegiea gigantea*; Ted Fleming, pers. comm., 2000); chainfruit cholla/choya *(Opuntia fulgida)* and royal poinciana/árbol del fuego (*Delonyx regis*; Eduardo Gómez L., pers. comm., 2001); ironwood (*Olneya tesota*; Reyna Castillo G., pers. comm., 2001); jackass clover (*Wislizenia refrecta*; Lin Piest, pers. comm., 2001); mezcal *(Agave angustifolia)* and sina (Francisco Molina F., pers. comm., 2001); and rama parda (*Ruellia californica*; Roberto Morales B., pers. comm., 2001). Alicia

Carrillo observed hummingbirds visiting an additional nine species of cultivated plants in her garden in Querobabi (see appendix 6.1). Including six other plants from miscellaneous sources, 101 species of plants are known to be visited by hummingbirds in Sonora (see appendix 6.1). We have also compiled a list of an additional thirty-one species of Sonoran plants reported to have been visited by hummingbirds in the United States or other Mexican states (see appendix 6.2). These numbers illustrate the adaptability of hummingbirds to a diversity of species and floral syndromes.

The diversity of hummingbirds visiting each of these plant species varied greatly. The plants attracting the most species of hummingbirds in Sonora were Texas betony (eleven), tree tobacco (ten), hierba del piojo (six), Madrean beardtongue (six), pineapple sage (six), tree morning glory (six), Indian paintbrush/periquito (*Castilleja patriotica*, five), limita (five), Maycoba sage (five), wild jícama (five), tree ocotillo (four), rama del toro (three), ocotillo (three), and uvalama (three).

Plants Utilized by Migrant Rufous Hummingbirds in Sonora

Rufous hummingbirds are generalists in their feeding strategy. We observed them visiting twenty-five species of plants (see appendix 6.3). Only the resident broad-billed hummingbird *(Cynanthus latirostris)* visited more plants (forty-seven species), though violet-crowned *(Amazilia violiceps)* and white-eared *(Hylocharis leucotis)* hummingbirds with nineteen and eighteen species, respectively, each had nearly as much dietary breadth. With the exception of chuparrosa and ocotillo (Calder 1993) and tree tobacco (Russell and Monson 1998), the plants that we observed rufous hummingbirds visiting are new records for Sonora. Rufous hummingbirds were previously reported to visit chuparrosa and Mohave beardtongue along the coast and in the Sierra Pinacate (Calder 1993). They have also been observed feeding at columnar cacti (cardón and sahuaro; Ted Fleming, pers. comm., 2000) flowers near Bahía Kino, and mezcal at Agualurca near Hermosillo (Franciso Molina F., pers. comm., 2001). Rufous hummingbirds have thus been observed feeding at flowers of a total of thirty plants in Sonora. Broad-billed hummingbirds are the most common urban hummingbirds in Hermosillo.

The flowers of the twenty-three native plants visited by rufous hummingbirds were red (52.2 percent), yellow (21.7 percent), white (13.0 percent), and purple (8.7 percent), indicating that they forage beyond floral choices representing the classic hummingbird flower syndrome. Rufous hummingbirds are associated with a flower color distribution similar to that of the eighty-four species of plants in Sonora visited by all hummingbirds: "reddish" (orange, red, pink, salmon, magenta; 49.4 percent), white (22.4 percent), yellow (16.5 percent), and lavender-purple (11.8 percent). With the exceptions of hierba del piojo (which has yellow flowers), tree morning glory (which has white flowers), and wild jícama (which has purple flowers), most of the large patches of plants frequented by hummingbirds in Sonora have red flowers (see appendix 6.3). We observed rufous as well as broad-billed, Costa's, and violet-crowned hummingbirds, bees, flies, and a monarch butterfly *(Danaus plexippus)* visiting the flowers of palo chino *(Havardia mexicana)* along the Río Yaqui near Tónichi. The flowers of this species are tiny white cups aggregated into dense heads. We speculate that hummingbirds may also visit some of the other common mimosoids, such as the acacias *(Acacia cochliacantha, A. greggii, A. occidentalis)*, mesquites *(Prosopis glandulosa, P. velutina)*, feather trees *(Lysiloma divaricatum, L. watsonii)*, and wait-a-minute bushes *(Mimosa aculeaticarpa, M. distachya,* and *M. palmeri)*. In these plants and many others, it is often difficult to see if hummingbirds are feeding on nectar or insects attracted to the flowers.

Rufous Hummingbird Competition for Nectar with Other Hummingbirds

In January 2001, we found that broad-billed and Costa's hummingbirds were widespread and very common in Sonoran desertscrub, foothills thornscrub, and tropical deciduous forest. They were defending territories established in stands of tree morning glories and tree tobaccos. A female broad-billed hummingbird was incubating eggs in a nest. Soon they had to defend their territories against rufous hummingbirds, which are notoriously aggressive. Rufous hummingbirds will dominate most other species reaching the United States, though they are subordinate to some tropical species in Mexico, including the larger magnificent *(Eugenes fulgens)* and the resident berylline *(Amazilia beryllina)*, violet-crowned, and white-eared hummingbirds in Jalisco (Des Granges 1979). Rufous hummingbirds were seen on

TABLE 6.1. Observations of Hummingbirds in the Same Area as Rufous Hummingbirds in Sonora, Dec. 1999–Sept. 2001

	No. of Observations	
Common Name	Spring	Summer
Anna's hummingbird	3	2
Berylline hummingbird	3	6
Black-chinned hummingbird	1	7
Blue-throated hummingbird	3	6
Broad-billed hummingbird	14	7
Broad-tailed hummingbird	2	5
Calliope hummingbird	1	0
Costa's hummingbird	14	3
Magnificent hummingbird	1	2
Violet-crowned hummingbird	4	6
White-eared hummingbird	2	8

the various trips to Sonora in 2000 and 2001 in the same areas as eleven other species of hummingbirds (table 6.1) and likely compete with some or all of them. The impact on resident hummingbird communities when rufous hummingbirds arrive in Sonora in February would be a very interesting study.

Predicting Spring Nectar Corridor Positions

Although the migration routes and stopover localities in Sonora for rufous hummingbirds remain poorly known, the nectar resources generally available to migrants passing through Sonora have well-defined geographic and seasonal distributions. The flower resources are potential nectar corridors that may or may not be used by hummingbirds but serve as hypotheses to be tested in the field and will help interpret migration patterns from field observations. It should be possible to deduce their routes from Jalisco northwestward through Sinaloa and Sonora to California, Washington, and Alaska based on phytogeographic and phenological patterns (Calder 1993). The tree morning glory is the only obvious plant with nectar available in

January and February that occurs continuously along the tropical coast of Mexico. From Nayarit to southern Sonora, its distribution follows the Sierra Madre Occidental, which edges the tropical lowlands into a narrow ribbon. In southern Sonora near the Sinaloa border (26°30' N), the coast becomes broader at the tropical zone between the Sierra and the Sea of Cortés, providing more favorable habitat for hummingbird plants. On the coastal plain, rainfall gradually decreases through the transition from coastal thornscrub to the Sonoran Desert near Guaymas (27°30' N; Turner and Brown 1982). North-bound rufous hummingbirds could potentially use one or more of three floral corridors (figure 6.2).

The foothills corridor is east of the Sonoran Desert through tropical deciduous forest and taller foothills thornscrub as far north as the Río Yaqui near Tónichi (28°34' N). This corridor largely follows the Río Mayo and Río Yaqui valleys northward until they veer eastward toward their headwaters in the barrancas in the Sierra Madre Occidental in Chihuahua. Tree ocotillos are more common in thornscrub in the lowlands near the rivers, and tree morning glories become common on the tropical slopes above. Both are locally common in patches almost to the U.S. border in the Río Sonora (30°11' N)/Río Moctezuma or the Río Fronteras/Río Bavispe (northern tributaries of the Río Yaqui) valleys.

The Plains of Sonora corridor passes through the Plains of Sonora subdivision of the Sonoran Desert through Hermosillo (29°05' N) to Santa Ana (30° N). From south to north, spring flowers are dependable on ocotillos, palo adán *(Fouquieria diguetii)*, and tree ocotillo as well as tree morning glories on emergent thornscrub-clad desert ranges. Rufous hummingbirds are regularly encountered from early March to early April in the Sonoran Desert north of Hermosillo in central Sonora, the Pinacate region of northwestern Sonora (Calder 1993), and in the Lower Colorado River Valley near El Golfo de Santa Clara in Sonora and Yuma in southwestern Arizona and adjacent California (Rosenberg et al. 1991; Lin Piest, pers. comm., 2001). By mid-April the leaders of the migration are becoming abundant near Juneau and coastal Alaska (Calder, pers. obs.).

Rufous hummingbirds using the Gulf Coast corridor following the coast through the Sonoran Desert to northwestern Sonora are feeding on palo adán, chuparrosa, and ocotillo. The central Gulf Coast subdivision of the Sonoran Desert is similar on both sides of the Sea of Cortés, and

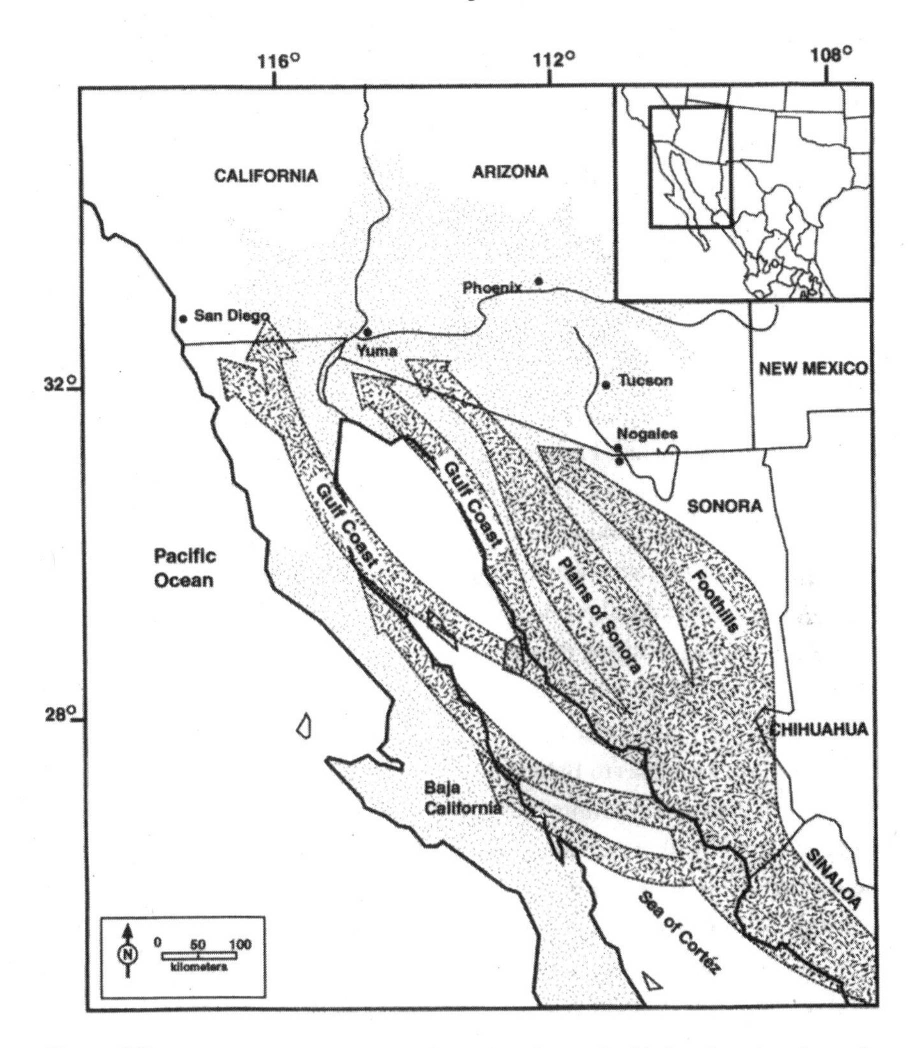

Figure 6.2. Potential nectar corridors for rufous hummingbirds migrating through Sonora, Mexico, in the spring.

hummingbirds could fly across the Gulf and continue northward through Baja California. In Sonora, we observed rufous on the coast, often in sparsely vegetated desertscrub in view of the water, at Bahía Kino (28°52' N), Desemboque del Río San Ignacio (ca. 29°25' N), Puerto Lobos (30°18' N), Desemboque del Río Concepción (30°34' N), Puerto Peñasco (31°18' N), and El Golfo de Santa Clara (31°30' N). Observations of rufous hummingbirds on Tiburón, San Esteban, San Lorenzo Sur, and Ángel de la Guarda (Cody

1983) suggest that some rufous island-hop across the Sea of Cortés, presumably reaching the coast of Baja California between 29° N and 30° N. Observations of large numbers of rufous and Allen's *(Selasphorus sasin)* hummingbirds at Mulegé (26°56' N) in March 1996 in Baja California Sur (Whitmore and Whitmore 1997) raise the possibility that some migrants may cross the Sea of Cortés much farther south, before reaching Guaymas (27°58' N) and the Sonoran Desert.

Tree morning glories and tree ocotillos provide the most dependable flowers each year. The northern end of the foothills corridor in the upper Río Sonora Valley near Arizpe (30°20' N) is too far east for most hummingbirds migrating to California. However, some rufous hummingbirds are seen in the spring in southeastern Arizona in unusual years like 2001.

Chuparrosa, palo adán, and ocotillo on the Gulf Coast corridor, which respond to variable winter-spring rainfall from Pacific frontal storms, are less dependable floral resources. The southern penetration of the storms is highly variable, resulting in regular severe droughts of several years south of Puerto Libertad (30° N) and in high dieback and mortality of shrubs, including chuparrosa. In many years, the southern portion of the Gulf Coast corridor from Guaymas to Bahía Kino is much drier than areas to the north. We observed rufous hummingbirds from the coast (112°50' W) east to El Aguajito (109°03' W, 1,640 m elevation; spring 2000) and Mesa del Campanero (109°01'40" W, 2,100 m elevation; spring 2001). Our observations suggest that all three nectar corridors were used in both years but to varying degrees depending on where rainfall from the previous season had generated the greatest floral abundances.

Predicting Summer Nectar Corridor Positions

In summer, the hummingbird migratory corridors are not so obvious. The details of the southerly migration are not well understood. There are fewer obvious flowers suitable for hummingbirds migrating through desertscrub and thornscrub late in the warm season. In thornscrub and tropical deciduous forest, a number of vines including queen's wreath/sanmiguelito (*Antigonon leptopus*, flowers pink) and various morning glories/trompillos have showy flowers (Convolvulaceae: *Merremia palmeri* [flowers white], *Operculina pteripes* [flowers salmon], *Ipomoea* spp. [flowers red, pink, blue, purple, white], and *Jacquemontia* spp. [flowers white]; Cucurbitaceae: *Schizocarpum palmeri* [flowers yellow]).

Above tropical deciduous forest in the Sierra Madre Occidental, summer floral abundance increases dramatically. Hierba del piojo is the most common flowering shrub in oak woodland that attracts hummingbirds in July and August. In pine-oak forest, hummingbirds concentrate in local dense patches where reddish flowers occur, including cardinal flower *(Lobelia cardinalis)*, india *(Zinnia peruviana)*, Indian paintbrush/periquito *(Castilleja patriotica, C. tenuiflora)*, Maycoba sage, pineapple sage, and Texas betony. In early September, rufous hummingbirds were aggressively defending territories in dense Indian paintbrush and Texas betony patches in the pine-oak forest zone at 2,100 m on Mesa del Campanero. We observed six to ten other species of hummingbirds, some of which were apparently excluded from the flowers by rufous hummingbirds.

In summer in the Sierra Madre Occidental, hummingbirds visit a diverse group of species that do not occur in dense patches but are widely scattered, such as bat-faced monkey flower/perrito colorado (*Cuphea llavea*, flowers red and purple), cigarrito (*Bouvardia ternifolia*, flowers red), Madrean beardtongue *(Penstemon wislizenii)*, and burgundy sage (*Salvia microphylla*, flowers magenta). Other species likely visited by hummingbirds include *Agastache mearnsii* (flowers magenta), beardtongues (*Penstemon campanulatus*, flowers purple), *Lamourouxia viscosa* (flowers pink), Madrean starthistle/cardo (*Centaurea rothrockii*), and other sages (*Salvia townsendii*, flowers red, and others). During this season, it is more difficult to observe hummingbirds feeding at these species because both the flowers and the birds are more dispersed and the nectar resources may not be sufficient for hummingbirds to establish territories.

ACKNOWLEDGMENTS

Lorene Calder, Mark A. Dimmitt, Reyna A. Castillo G., Lucy Daley, George Ferguson, Cam and Joy Finlay, Joan Day-Martin, Francisco Molina F., Lee Rogers, and Susan Wethington helped with the fieldwork. The teachers Rogelio Amaya C., Merardo Cano R., and Rogelio Martínez B. and the Seri Indian Para-Ecólogos Roberto Molina R., Fernando Morales O., and Héctor Perales M. provided plant phenological data and hummingbird observations. Castillo, Molina, Ted Fleming, and Eduardo Gómez L. provided important additional observations of hummingbirds visiting Sonoran Desert plants. Molina sponsored the field activities under his national collecting permit. Field activities were supported by a grant from the Turner Foundation to the Arizona-Sonora Desert Museum. Jack Whetstone provided information on the hummingbirds observed on the Yécora Christmas Bird Count. Richard Erickson provided information on hummingbirds in Baja California. Don Rubén Coronado and Doña Ubelina Quijada, and their sons Oscar and Sergio Coronado, enthusiastically adopted hummingbird feeders at Restaurant La Palmita. The manuscript was improved by the editorial efforts of Susan Wethington, Mark Dimmitt, Tani Hubbard, and Gary P. Nabhan. Gerald Dewavendewa drafted the maps.

Appendix 6.1. Plants Visited by Hummingbirds in Sonora

SOURCES

[1] Observations of the authors in 2000–2001 and Yécora Christmas Bird Count (Jack Whetstone, pers. comm., 2000)
[2] Gentry 1942
[3] Marshall 1957
[4] Calder 1993
[5] Baltosser and Scott 1996
[6] Russell and Monson 1998
[7] Powers and Wethington 1999
[8] Johnsgard 1997
[9] Ted Fleming, pers. comm., 2000
[10] Francisco Molina F., pers. comm., 2001
[11] Reyna Castillo G., pers. comm., 2001
[12] Eduardo Gómez L., pers. comm., 2001
[13] Roberto Morales B., pers. comm., 2001
[14] Alicia Carrillo, pers. comm., 2001
[15] Lin Piest, pers. comm., 2001.

Flower color and flowering times from Martin et al. 1998, Van Devender and Reina G.'s personal collection database, and data from herbarium specimen labels

ABBREVIATIONS

ALHU Allen's hummingbird, *Selasphorus sasin*
ANHU Anna's hummingbird, *Calypte anna*
BBLH broad-billed hummingbird, *Cynanthus latirostris*
BCHU black-chinned hummingbird, *Archilochus alexandri*

BEHU berylline hummingbird, *Amazilia beryllina*
BLHU bumblebee hummingbird, *Selasphorus heliosa*
BLUH blue-throated hummingbird, *Lampornis clemenciae*
BTLH broad-tailed hummingbird, *Selasphorus platycercus*
CAHU calliope hummingbird, *Stellula calliope*
CIHU cinnamon hummingbird, *Amazilia rutila*
COHU Costa's hummingbird, *Calypte costae*
LUHU Lucifer hummingbird, *Calothorax lucifer*
MAHU magnificent hummingbird, *Eugenes fulgens*
PCST plain-capped Starthroat, *Heliomaster constantii*
RUHU rufous hummingbird, *Selasphorus rufus*
VCHU violet-crowned hummingbird, *Amazilia violiceps*
WEHU white-eared hummingbird, *Hylocharis leucotis*
XAHU xantus hummingbird, *Hylocharis xantusii*

Acanthaceae

Anisacanthus andersonii. Limita. Shrub. Red; February–April. BBLH, BEHU, RUHU, VCHU, WEHU [1]

Anisacanthus thurberi. Desert honeysuckle. Shrub. Red-orange to orange; January–April. COHU, RUHU [1, 2]

Justicia californica. Chuparrosa. Shrub. Red to orange; (October) February–April. COHU [1], RUHU [4]

Justicia candicans. Rama del toro/chuparrosa. Subshrub. Red; November–April. BBLH, COHU, WEHU [1]

Ruellia californica. Rama parda. Subshrub. Purple; December–March (September). COHU [14]

Tetramerium abditum. Rama del toro. Subshrub. Red-orange; (January) February–April (May). BBLH [1]

Agavaceae

Agave angustifolia. Agave; mezcal, maguey. Rosette succulent. Yellow; March. COHU, RUHU [11]

Agave palmeri. Agave; maguey. Rosette succulent. Reddish to yellowish; July–August. BCHU [1]

Agave shrevei. Agave; lechuguilla ceniza. Rosette succulent. Light green to pale yellow; August–September. BCHU, COHU, RUHU, WEHU [1]

Agave vivipara (= *A. angustifolia*). Agave; mezcal, maguey. Rosette succulent. Yellow; March. COHU, RUHU [11]

Agave sp. Agave/maguey. Rosette succulent. Yellow. LUHU [7], MAHU [3]

Apocynaceae

Mandevilla foliosa. Hierba del piojo. Subshrub. Yellow; June–August. BBLH, BCHU, BEHU, RUHU, VCHU, WEHU [1, 7]

Nerium oleander. Oleander; laurel. Cultivated shrub. White or pink; February–October. BBLH, VCHU [1]

Thevetia peruviana. Yellow oleander. Cultivated shrub. Peach; March–May. BBLH [1]

Asclepiadaceae

Asclepias curassavica. Bloodflower milkweed; hierba de la cucaracha. Cultivated perennial herb. Red; December–May, August. BBLH [1]

Balsaminaceae

Impatiens balsamina. Impatiens. Cultivated herb. Variable colors; September. MAHU [1]

Bignoniaceae

Campsis radicans. Trumpet creeper; corneta. Cultivated perennial vine. Red; July–August. BBLH, BCHU [1]

Tabebuia impetiginosa. Amapa. Pink-purple; January–April. BBLH, VCHU [1]

Tecomaria capensis. Cape honeysuckle; bignonia. Cultivated perennial herb. Red-orange; fall, spring. BBLH [1]

Tecoma stans. Trumpet flower; gloria, lluvia de oro, palo de arco. Cultivated tree/shrub. Yellow. BBLH [1]

Cactaceae

Carnegiea gigantea. Sahuaro. Tree cactus. White; April–May. BBLH, COHU, RUHU [10]

Echinocereus sp. Hedgehog cactus/choyita. Claret cup cactus. Red or red-orange; May. BTLH [3]

Opuntia [Cylindropuntia] fulgida. Chainfruit cholla; choya. Stem succulent. Rose; (April) May–August. COHU [13]

Opuntia [Cylindropuntia] thurberi. Cholla; siviri. Stem succulent. Yellow; January–May. COHU [1]

Opuntia [Nopalea] karwinskiana. Red prickly pear; nopal. Cultivated prickly pear. Red; February–March. BBLH, VCHU [1]

Opuntia [Platyopuntia] ficus-indica. Tree prickly pear; nopal de castilla/tuna. Cultivated prickly pear. Yellow; April–May. BBLH [1]

Opuntia [Platyopuntia] gosseliniana. Purple prickly pear; duraznilla. Prickly pear. Yellow; April–June. BBLH [1]

Pachycereus pecten-aboriginum. Etcho. Tree cactus. White; December–April. [2]

Pachycereus pringlei. Cardón. Tree cactus. White; April–June. BBLH, COHU, RUHU [9]

Stenocereus alamosensis. Galloping cactus; sina. Shrub cactus. Red; March–May. BBLH, COHU [1, 10]

Stenocereus thurberi. Organpipe cactus; pitahaya. Tree/shrub cactus. Red outside, white inside; (March) April–July (December). BBLH, COHU, RUHU [1, 9]

Campanulaceae

Lobelia cardinalis. Cardinal flower. Perennial herb. Red; April–September. RUHU, WEHU [1]
Lobelia laxiflora. Bellflower. Perennial herb. Red and yellow; March–June. BBLH, BTLH, RUHU [1], BLUH, BTLH, MAHU [3]

Capparaceae

Cleome isomeris [Isomeris arborea]. Bladderpod. Yellow; February–March (fall). ANHU [1]
Wislizenia refracta. Jackass clover. Yellow. RUHU [15]

Caprifoliaceae

Sambucus nigra subsp. *cerulea*. Mexican elderberry, sauco. Native tree, often cultivated. White; warm season. [14]

Compositae

Bidens sambucifolia. Tostón. Suffrutescent herb. Orange; September–March. BBLH [1]
Centaurea rothrockii. Madrean starthistle; cardo. Perennial herb. Purple and white; August–September. BLUH, RUHU, WEHU [1]
Cosmos sulfureus. Tostón. Native suffrutescent herb, often cultivated. Orange; September–December. [14]
Dianthus caryophyllus. Carnation; clavel. Cultivated herb. Red; September. MAHU [1]
Stevia plummerae. Perennial herb/subshrub. White or pinkish purple; July–October. BEHU [1]
Verbesina felgeri. Shrub. Yellow; various times. ANHU, BBLH [1]
Zinnia peruviana. Wild zinnia; india. Annual herb. Red; July–October. BBLH, BEHU, VCHU [1]

Convolvulaceae

Ipomoea arborescens. Tree morning glory; palo blanco/santo. Tree. White; (October) November–March (April). ANHU, BBLH, BEHU, COHU, PCST, VCHU, WEHU [1]
Ipomoea [Exogonium] bracteata. Wild jícama. Woody vine. Purple, pink, or fuchsia; (December) January–May. BBLH, BEHU, RUHU [1]
Ipomoea parasitica. Morning glory, trompillo. Herbaceous vine. Purple; September. BBLH [1]
Merremia palmeri. Morning glory vine; trompillo. Perennial vine. White; most months. BBLH, COHU [1]
Operculina pteripes. Morning glory vine; trompillo. Perennial vine. Salmon; July–October. BBLH [1]

Ericaceae

Arbutus xalapensis. Madrone; madroño. Tree. Greenish white; March–April. BLUH, WEHU [1]
Arctostaphylos pungens. Manzanita; manzanilla. Shrub. White with pink base; March (May). BTLH [3], WEHU [1]

Euphorbiaceae

Ricinus communis. Castor bean, higuerilla. Herb, shrub, or tree. Red; July. BBLH [1]

Fouquieriaceae

Fouquieria diguetii. Palo adán. Tree/shrub. Red; February–May. BBLH, RUHU [1], COHU [1, 5]
Fouquieria macdougalii. Tree ocotillo. Tree/shrub. Red; February–March, July–October (rest of year). ANHU, BBLH, COHU, RUHU, VCHU [1], COHU [5]
Fouquieria splendens. Ocotillo. Shrub. Red; March–May (October). BBLH, COHU, RUHU [1, 4], BBLH [9], COHU [6]

Grossulariaceae (Saxifragaceae)

Ribes dugesii. Gooseberry. Shrub. Purple; February–March. WEHU [1]

Iridaceae

Gladiolus sp. Gladiola. Cultivated perennial herb. Red; August–September. WEHU [1]

Labiatae

Monarda citriodora var. *austromontana*. Bee balm; orégano del buro. Perennial herb. White with purple tint; July–October. BCHU [1]
Salvia betulaefolia. Maycoba sage. Shrub. Red; August–September. BCHU [1]
Salvia elegans. Pineapple sage. Subshrub. Red; March–October. BBLH, BCHU, BEHU, RUHU, VCHU, WEHU [1]
Salvia microphylla. Burgundy sage. Subshrub. Magenta; August–November. BCHU, BEHU [1]
Stachys coccinea. Texas betony. Herbaceous perennial. Red; March–October. ANHU, BBLH, BCHU, BEHU, BLUH, BTLH, COHU, MAHU, RUHU, VCHU, WEHU [1]

Leguminosae: Caesalpinioideae

Bauhinia variegata. Orchid tree; orquídea. Cultivated tree. Purple; mid-winter–spring. BBLH [1]
Caesalpinia pulcherrima. Mexican red bird-of-paradise; tabachín. Shrub. Red and yellow; March–September. BBLH [1, 2, 7, 8]
Delonyx regis. Royal poinciana; árbol del fuego, tabachín. Cultivated tree. Red; warm season. ANHU, BBLH, BCHU, COHU [12]
Haematoxylum brasiletto. Brasil. Tree. Yellow; April–October (primarily summer). BBLH [1]

Leguminosae: Mimosoideae

Havardia [Pithecellobium] mexicana. Palo chino. Tree. White; January–April. BBLH, COHU, RUHU, VCHU [1]

Parkinsonia microphylla. Foothills paloverde. Tree. Yellow; (February) March–May. COHU [5, 11]
Prosopis glandulosa. Honey mesquite; mezquite. Tree/shrub. Yellow; March–August. COHU [1, 9], RUHU [1]

Leguminosae: Papilionoideae

Cologania cf. *angustifolia*. Pea vine; frijol. Annual vine. Pink; June–September. BEHU [1]
Coursetia glandulosa. Samo, sámota. Shrub/tree. White, yellow, and pink; January–May. BBLH [1]
Olneya tesota. Ironwood; palo fierro. Tree/shrub. Lavender to purple; (March) April–May (June). COHU [9]
Robinia neomexicana. New Mexican locust. Shrub. White to purple; May–July. BTLH [3]

Liliaceae

Canna indica. Canna lily; perico sencillo, cuenta. Cultivated perennial herb. Red; March. BBLH [1]
Hemerocallis fulva. Lily; lirio. Cultivated perennial herb. Yellow; September. [14]

Lythraceae

Cuphea llavea. Bat-faced monkeyflower; perrito colorado. Perennial herb. Red and purple; (June) August–September (November). BEHU, VCHU [1]

Malpighiaceae

Callaeum [Mascagnia] macropterum. Gallinitas, batanene. Woody vine. Yellow; most of the year. BBLH, COHU [1]

Malvaceae

Hibiscus rosa-sinensis. Hibiscus; obelisco. Cultivated shrub. Red; mid-spring until fall. BBLH [1]
Hibiscus syriacus. Bibisca, obelisco. Cultivated shrub. Lavender, pink; August–September. BCHU [1]
Kosteletzkya thurberi. Obelisco. Shrub. Pink; September–October. BLHU [1]

Musaceae

Musa paradisiaca. Banana, plátano. Giant cultivated herb. Red. [14]

Myrtaceae

Eucalyptus globosus. Eucalyptus, eucalypto. Cultivated tree. Whitish; October. ANHU [1, 8]
Lagerstroemia indica. Crepe myrtle, atmosférica. Cultivated shrub. Lavender; various. [14]

Nyctaginaceae

Bougainvillea spectabilis. Bougainvillea; bugambilia. Cultivated woody vine. Magenta, white, various; February–April. ANHU, BBLH, VCHU [1]

Polemoniaceae

Ipomopsis thurberi. Purple starflower. Perennial herb. Purple; August–October. RUHU [1]

Polygonaceae

Antigonon leptopus. Queen's wreath; sanmiguelito. Perennial vine. Pink, red; August–April. BBLH, RUHU [1]

Rosaceae

Punica granatum. Pomegranate, granada. Cultivated shrub. Red. [14].
Rosa sp. Rose; rosal. Cultivated shrub. Pink; spring and later. ANHU, BBLH, BCHU, MAHU, WEHU [1]

Rubiaceae

Bouvardia ternifolia. Scarlet bouvardia; cigarrito, chuparrosa. Shrub. Red. RUHU, VCHU [1, 2]

Scrophulariaceae

Antirrhinum majus. Snapdragon, perritos. Cultivated annual herb. Yellow. BEHU, WEHU [1]
Castilleja patriotica. Indian paintbrush; periquito, perico. Parasitic perennial herb. Red-orange and yellow; July–October. BCHU, BTLH, RUHU, VCHU, WEHU [1]
Castilleja tenuiflora. Indian paintbrush; periquito, perico. Parasitic perennial herb. Red or red-orange and yellow; February–October. WEHU [1]
Penstemon parryi. Parry penstemon; varita de San José, pichelitos. Perennial herb. Pink-purple; spring. BBLH [1], COHU [5]
Penstemon pseudospectabilis. Mohave beardtongue. Perennial herb. Deep pink to rose-purple; February–May. COHU, RUHU [4]
Tecoma stans var. *stans*. Yellow trumpet bush, lluvia de oro, caballito, gloria. Cultivated shrub/tree. Yellow; warm part of year. BBLH [1]

Solanaceae

Lycium andersonii. Wolfberry; cacaculo, manzanita. Shrub. White; February–April, August–October. BBLH [1]
Lycium berlandieri. Wolfberry; huichutilla. Shrub. White or lavender; February–November. COHU [5]

Nicotiana glauca. Tree tobacco; juanloco. Introduced shrub. Yellow-green; all year. ANHU, BBLH, BCHU, BEHU, BLUH, COHU, MAHU, RUHU, VCHU, WEHU [1], BBLH, RUHU [6], BLUH, MAHU [3]
Petunia violacea. Petunia. Cultivated herb. Various; warm part of year. [14]
Solanum seaforthianum. Bellísima. Cultivated perennial vine. Purple or white; most of the year. [14]

Tropaeoliaceae

Tropaeolium majus. Nasturtium; mastuerzo. Cultivated herb. Yellow or orange. March, September. MAHU, RUHU [1]

Verbenaceae

Lantana camara. Lantana, confituría negra. Native shrub widely cultivated. Yellow, red-orange, pink. [14]
Vitex mollis. Uvalama. Tree. Lavender; March–July. BBLH, BCHU, COHU [1]

Appendix 6.2. Additional Hummingbird Plants Likely to Be Used Elsewhere in Sonora, Based on Feeding Observations in the Southwestern United States or Farther South in Mexico

SOURCES

[1] Baltosser and Scott 1996
[2] Johnsgard 1997
[3] Powers and Wethington 1999
[4] Scott 1994
[5] Calder and Calder 1992
[6] Powers 1996
[7] Calder 1993
[8] Arizmendi and Ornelas 1990
[9] Des Granges 1979
[10] Arriaga et al. 1990
[11] Lyon 1976
[12] Crosswhite and Crosswhite 1981

Note: Abbreviations for hummingbirds are in appendix 6.1.

Agavaceae

Agave deserti. Desert agave; maguey. Yellow. COHU [1]
Agave parryi. Parry agave. Rosette succulent; maguey. Yellow. BBLH [3], MAHU [2]
Agave schottii. Shindagger; amole. Rosette succulent. Yellow. BBLH [3]

Asclepiadaceae

Asclepias subulata. Desert milkweed; candelilla. Shrub. White. COHU [1]

Bignoniaceae

Chilopsis linearis. Desert willow; mimbre. Tree. White, pink. BBLH [3], COHU [1], LUHU [4]

Cactaceae

Echinocereus coccineus. Claret cup cactus. Hedgehog cactus. Red. BTLH [5, as *E. triglochidiatus*]
Echinocereus polyacanthus. Claret cup cactus. Hedgehog cactus. Red-orange. MAHU [6]
Opuntia acanthocarpa. Buckhorn cholla; siviri. Stem succulent. Yellow, orange, red. COHU [1]

Caprifoliaceae

Lonicera involucrata. Honeysuckle. MAHU [2]

Carophyllaceae

Silene laciniata. Indian pink. Perennial herb. Red. ANHU [2], BTLH [5], MAHU [6]

Compositae

Cirsium neomexicanum. New Mexican thistle; cardo. Perennial herb. Purple. BBLH [3]

Convolvulaceae

Ipomoea quamoclit. Scarlet quamoclit; trompillo. Annual vine. Red. CIHU [8]

Hydrophyllaceae

Wigandia urens. Shrub. White. CAHU [9]

Labiatae

Hyptis albida (= *H. emoryi*). Desert lavender; salvia. Shrub. Purple to blue-purple. COHU [1]
Hyptis mutabilis. Shrub. Pale lavender to purple. [9]
Leonotis nepetaefolia. Lion's ear; cordoncillo de San Francisco. Introduced annual herb. Orange. MAHU [6]
Salvia lemmonii. Salvia; chía. Perennial herb. Lavender-pink or crimson. BBLH [3]

Leguminosae: Mimosoideae

Chloroleucon mangense (Pithecellobium undulatum). Ape's earring; palo fierro/pinto. Tree. White. BBLH [3, 8]

Leguminosae: Papilionoideae

Erythrina flabelliformis. Coral bean; chilicote. Red. [2]

Loasaceae

Eucnide cordata. Rock nettle; pega-pega. Perennial herb. Cream. COHU [1]

Loranthaceae

Psittacanthus calycosus. Mistletoe/toji. Woody parasite. Orange. BEHU, CIHU, VCHU [9]

Nyctaginaceae

Mirabilis jalapa. Four o'clock; maravilla. Subshrub. Magenta. BBLH [3], LUHU [4], XAHU [10]

Onagraceae

Epilobium canum (= *Zauschneria californica*). Wild fuchsia. Perennial herb. Red. ALHU [2], BBLH [3]

Ranunculaceae

Delphinium scaposum. Barestem larkspur; espuelita de campo. Perennial herb. Lavender to purple to blue. COHU [1]

Scrophulariaceae

Lamourouxia viscosa. Perennial herb. Pink. BLHU, BLUH, MAHU [6, 11]
Mimulus cardinalis var. *verbenaceus*. Crimson monkeyflower. Perennial herb. Red-orange. [2]
Penstemon barbatus. Scarlet bugler. Perennial herb. Red. BBLH [3], BTLH, RUHU [5, 7, 12]

Solanaceae

Lycium exsertum. Wolfberry. Shrub. White. COHU [1]
Nicotiana obtusifolia (= *N. trigonophylla*). Wild tobacco. Perennial/annual herb. Greenish yellow. [12]

Verbenaceae

Lippia umbellata. Bacatón. Shrub. Yellow. BTLH, MAHU [3, 6, 9]

Zygophyllaceae

Larrea divaricata. Creosotebush; gobernadora, hediondilla. Shrub. COHU [1]

Appendix 6.3. Observations of Rufous Hummingbirds Visiting Plants in Sonora

Note: $n = 30$

CODES

*	important nectar source
+	introduced
!	cultivated

ABBREVIATIONS

Life forms

CC	columnar cactus
HP	herbaceous perennial
PV	perennial vine
RS	rosette succulent
SH	shrub
SS	subshrub
TR	tree

Flower colors

O	orange
OR	orange-red
PU	purple
PW	purple and white
R	red
RP	red or pink
RY	red and yellow
W	white
Y	yellow
YG	yellow-green

Seasons

SP	spring
SU	summer

SOURCES

[1] This study
[2] Ted Fleming, pers. comm., 2001
[3] Calder 1993
[4] Russell and Monson 1998
[5] Francisco Molina F., pers. comm., 2001
[6] Lin Piest, pers. comm., 2001

Species	Common Name	Life Form/ Color	No. of Observations/ Season
Beardtongue [3]	*Penstemon pseudospectabilis*	HP-R	[?]-SP
Bellflower [1]	*Lobelia laxiflora*	HP-R	1-SP
! Bougainvillea [1]	*Bougainvillea spectabilis*	PV-PW	2-SP
! Cape honeysuckle [1]	*Tecomaria capensis*	HP-OR	1-SP
Cardinal flower [1]	*Lobelia cardinalis*	HP-R	1-SU
Cardón [2]	*Pachycereus pringlei*	CC-W	[?]-SP
Chuparrosa [3]	*Justicia californica*	SH-OR	7-SP
Cigarrito [1]	*Bouvardia ternifolia*	SH-R	1-SU
* Desert honeysuckle [1]	*Anisacanthus thurberi*	SH-OR	1-SP
* Hierba del piojo [1]	*Mandevilla foliosa*	SH-Y	1-SP
Honey mesquite [1]	*Prosopis glandulosa*	TR-Y	1-SP
* Indian paintbrush [1]	*Castilleja patriotica*	HP-OR	1-SU
Jackass clover [6]	*Wislizenia refracta*	SS-Y	1-SP
Lechuguilla [1]	*Agave shrevei*	SR-YG	1-SU
* Limita [1]	*Anisacanthus andersonii*	SH-R	1-SP
Madrean starthistle [1]	*Centaurea rothrockii*	HP-PW	1-SU
Mezcal [5]	*Agave angustifolia*	RS-Y	2-SP
! Nasturtium [1]	*Tropaeolium majus*	HP-O	1-SP
Ocotillo [1, 3]	*Fouquieria splendens*	SH-R	5-SP
Palo adán [1]	*Fouquieria diguetii*	TR-R	2-SP
Palo chino [1]	*Havardia mexicana*	TR-W	1-SP
* Pineapple sage [1]	*Salvia elegans*	SS-R	4-SU
Purple starflower [1]	*Ipomopsis thurberi*	HP-PU	1-SU
Queen's wreath [1]	*Antigonon leptopus*	PV-R	1-SU
Red bird-of-paradise [1]	*Caesalpinia pulcherrima*	SH-RY	1-SU
Sahuaro [2]	*Carnegiea gigantea*	CC-W	SP
Texas betony [1]	*Stachys coccinea*	HP-RP	3-SU
* Tree ocotillo [1]	*Fouquieria macdougalii*	TR-R	14-SP
+ Tree tobacco [1, 4]	*Nicotiana glauca*	SH-YG	2-SU
* Wild jícama [1]	*Ipomoea bracteata*	PV-PU	2-SP

LITERATURE CITED

Arizmendi, M. C., and J. F. Ornelas. 1990. Hummingbirds and their floral resources in a tropical dry forest in Mexico. Biotropica 22:172–80.

Arriaga, L., R. Ricardo E., and A. Ortega R. 1990. Endemic hummingbirds and madrones of Baja California: Are they mutually dependent? The Southwestern Naturalist 35:76–79.

Baltosser, W. H., and P. E. Scott. 1996. Costa's hummingbird *(Calypte costae)*. Pp. 1–31 in A. Poole and F. Gill, eds., The Birds of North America, No. 251. The Academy of Natural Sciences, Philadelphia; The American Ornithologists' Union, Washington, D.C.

Búrquez M., A., A. Martínez Y., and P. S. Martin. 1992. From the high Sierra Madre to the coast: Changes in vegetation along highway 16, Maycoba-Hermosillo. Pp. 239–52 in K. F. Clark, J. Roldán, and R. H. Schmidt, eds., Geology and Mineral Resources of the Northern Sierra Madre Occidental. Guidebook, El Paso Geological Survey Publication No. 24, El Paso.

Calder, W. A. 1993. Rufous hummingbird *(Selasphorus rufus)*. Pp. 1–18 in A. Poole and F. Gill, eds., The Birds of North America, No. 53. The Academy of Natural Sciences, Philadelphia; The American Ornithologists' Union, Washington, D.C.

Calder, W. A., and L. L. Calder. 1992. Broad-tailed hummingbird *(Selasphorus platycercus)*. Pp. 1–15 in A. Poole and F. Gill, eds., The Birds of North America, No. 16. The Academy of Natural Sciences, Philadelphia; The American Ornithologists' Union, Washington, D.C.

Cody, M. L. 1983. The land birds. Pp. 210–45 in T. J. Case and M. L. Cody, eds., Island Biogeography in the Sea of Cortés. University of California Press, Berkeley.

Crosswhite, F. S., and C. D. Crosswhite. 1981. Hummingbird pollinators of flowers in the red-yellow segment of the color spectrum, with special reference to *Penstemon* and the "open habitat." Desert Plants 3:156–70.

Des Granges, J. 1979. Organization of a tropical nectar feeding bird guild in a variable tropical environment. The Living Bird 17:199–236.

Gentry, H. S. 1942. Río Mayo Plants: A Study of the Río Mayo, Sonora. Carnegie Institution of Washington Publication 527, Washington, D.C.

Johnsgard, P. A. 1997. The Hummingbirds of North America. 2d ed. Smithsonian Institution Press, Washington, D.C.

Lyon, D. L. 1976. A montane hummingbird territorial system in Oaxaca, Mexico. The Wilson Bulletin 88:280–99.

Marshall, J. T. 1957. Birds of the pine-oak woodland in southern Arizona and adjacent Mexico. Cooper Ornithological Society. Pacific Avifauna (32):1–125.

Martin, P. S., D. Yetman, M. Fishbein, P. Jenkins, T. R. Van Devender, and R. K. Wilson. 1998. Gentry's Río Mayo Plants. The Tropical Deciduous Forest and Environs of Northwest Mexico. University of Arizona Press, Tucson.

Phillips, A. R. 1975. The migrations of Allen's and other hummingbirds. The Condor 77:196–205.

Powers, D. R. 1996. Magnificent hummingbird. Pp. 1–19 in A. Poole and F. Gill, eds., The Birds of North America, No. 221. The Academy of Natural Sciences, Philadelphia; The American Ornithologists' Union, Washington, D.C.

Powers, D. R., and S. M. Wethington. 1999. Broad-billed hummingbird. Pp. 1–17 in A. Poole and F. Gill, eds., The Birds of North America, No. 430. The Academy of Natural Sciences, Philadelphia; The American Ornithologists' Union, Washington, D.C.

Reina G., A. L., T. R. Van Devender, W. Trauba, and A. Búrquez M. 1999. Caminos de Yécora. Notes on the Vegetation and Flora of Yécora, Sonora. Pp. 137–44 in D. Vásquez del Castillo, M. Ortega N., and C. A. Yocupicio C., eds., Symposium Internacional sobre la Utilización y Aprovechamiento de la Flora Silvestre de Zonas Arida, Universidad de Sonora, Hermosillo.

Rosenberg, K. V., R. D. Ohmart, W. C. Hunter, and B. C. Anderson. 1991. Birds of the Lower Colorado River Valley. University of Arizona Press, Tucson.

Russell, S. M., and G. Monson. 1998. The Birds of Sonora. University of Arizona Press, Tucson.

Scott, P. E. 1994. Lucifer hummingbird *(Calothorax lucifer)*. Pp. 1–19 in A. Poole and F. Gill, eds., The Birds of North America, No. 13. The Academy of Natural Sciences, Philadelphia; The American Ornithologists' Union, Washington, D.C.

Turner, R. M., and D. E. Brown. 1982. Sonoran desertscrub. Desert Plants 4:121–81.

Waser, N. M. 1979. Pollinator availability as a determinant of flowering times in ocotillo *(Fouquieria splendens)*. Oecologia 39:107–21.

Whitmore, R. C., and R. C. Whitmore. 1997. Late fall and early spring bird observations for Mulegé, Baja California Sur, Mexico. Great Basin Naturalist 57:131–41.

CHAPTER 7

Saguaros and White-Winged Doves

The Natural History of an Uneasy Partnership

CARLOS MARTÍNEZ DEL RIO, BLAIR O. WOLF,
AND RUSSELL A. HAUGHEY

The western white-winged dove *(Zenaida asiatica mearnsii)* is unique among migrant pollinators because it is also a game species. These doves pollinate saguaro flowers and consume saguaro fruit, and they are also intensely hunted in their breeding range in Sonora and Arizona, and in their wintering range in western Mexico. The flowers of the giant columnar saguaro cactus *(Carnegiea gigantea)* are pollinated by these birds. The conspicuous interplay between saguaros and white-winged doves has become a prominent feature of the iconography of the natural history of the Sonoran Desert. Images of doves plunging their heads into saguaro flowers and messily eating the bright red pulp of saguaro fruit adorn postcards, magazine covers, and even children's books (Weidner-Zoehfeld and Mirocha 1997). These images fail to illustrate the intricacies and ambiguities in the interactions between these two species, or the unique characteristics of these interactions in the special context of the phenomenon of pollination by migratory animals in the Sonoran Desert.

Saguaros

Saguaros are among the most charismatic members of the diverse Sonoran Desert flora. Not only are these large columnar cacti regional icons in the United States, but they also play a crucial role in the energy and water economy of a large number of animal species (Shreve 1945, 1951). In an ecosystem distinguished by unpredictability and scarcity of water and nutrients, saguaros are abundant and reliable providers of these resources (Thackery and Leding 1929; Steenbergh and Lowe 1977). Saguaros can live for 160 years or more; healthy plants begin to reproduce when they are 30–40 years old and about 2 m tall (Steenbergh and Lowe 1977; Pierson and Turner 1998). Older plants can produce up to 500 flowers each season. Although there is considerable variation in the timing of flower and fruit production (Bowers 1996), year-to-year productivity of saguaro stands is remarkably consistent (Schmidt and Buchmann 1986; Fleming et al. 1996). Even long periods of drought do not appear to appreciably affect the fruit and flower output of a stand (Thackery and Leding 1929; Alcorn et al. 1959; McGregor et al. 1962; Steenbergh and Lowe 1977; Haughey 1986).

Saguaros produce flowers in the late spring (from late April to mid-June in Arizona; Schmidt and Buchmann 1986) and their fruit ripens from mid-June through August (Steenbergh and Lowe 1977). In good years, saguaro seeds are dispersed (often defecated by an animal) just prior to the monsoon rains, which provide conditions that favor germination and establishment. Because the reproduction of saguaros overlaps with the driest and often hottest months of the year, their flowers and fruit may represent crucial water and nutrient sources for a variety of desert animals. Saguaro flowers and fruit are used by almost every member of the bird community in the Sonoran Desert (Steenbergh and Lowe 1977; Haughey 1986).

Saguaro flowers are self-sterile and bloom for a single evening. They typically open between ten o'clock and midnight and close at about noon of the following day (Peebles and Parker 1946; Fleming et al. 1996). Nectar production starts after the flowers open and continues all night and for several hours after sunrise. Each flower produces about 2 ml of nectar and 0.5 g of sugar (Fleming et al. 1996). In addition, flowers produce large amounts of pollen (ca. 300 mg per flower; Schmidt and Buchmann 1986). Nectar-feeding bats *(Leptonycteris curasoae* and *Choeronycteris mexicana)* feed on saguaro nectar and pollen at night. In Arizona, several bird species visit saguaro flowers during the morning (*Zenaida asiatica, Melanerpes uropygialis, Toxostoma curvirostre*, and *Icterus parisorum*, among others; see Fleming et al. 1996). Farther south in Sonora's coastal desert, where saguaros bloom earlier, migrant warblers (*Vermivora celata, V. ruficapilla, Dendroica petechia*, and *Willsonia pusilla*) feed on the saguaro's nectar during their northward migration (T. Fleming, personal communication). Introduced honeybees *(Apis mellifera)* and native bees (several species of *Diadasia* spp. and *Bombus* spp.; S. Buchmann, pers. comm.) also intensely visit saguaro flowers and collect nectar and pollen (Alcorn et al. 1959; Schmidt and Buchmann 1986).

In Arizona, saguaro fruits ripen from mid-June through August. Peak production occurs in early to mid-July. Ripe fruits split open, forming an open goblet that exposes a mass of deep red pulp peppered with shiny black seeds. When fresh, the juicy fruit matrix contains large amounts of water (82 percent of the total mass) and sugar (92 percent of the dry pulp) and a small amount of protein (4 percent dry weight). Curiously, saguaro fruit pulp has a remarkably high calcium content (Ross 1944; Greenhouse 1979). The tiny saguaro seeds imbedded in the pulp are nutritious and numerous. Although each seed is very small (1.3 mg; Steenbergh and Lowe 1977), the total mass of seeds represents 40–45 percent of the 6.6 g of dry pulp contained in each fruit. For animals that can break the seed coat, seeds represent a rich source of lipid (ca. 30 percent dry mass), protein (ca. 16 percent dry mass), and carbohydrate (54 percent dry mass; Greenhouse 1979).

Bats *(L. curasoae* and *C. mexicana)* and a large number of avian species feed on saguaro fruit as it opens. A partial list of birds known to feed on saguaro fruit in Arizona and Sonora, Mexico, includes Gambel's quail *(Callipepla gambelii)*, white-winged dove *(Zenaida asiatica)*, Gila woodpecker *(Melanerpes uropygialis)*, ash-throated flycatcher *(Myiarchus cinerascens)*, verdin *(Auriparus flaviceps)*, curve-billed thrasher *(Toxostoma curvirostre)*, black-headed grosbeak *(Pheucticus melanocephalus)*, northern cardinal *(Cardinalis cardinalis)*, and house finch (*Carpodacus mexicanus*; Sosa 1997; Wolf and Martínez del Rio 2000). At the beginning and end of the season, when ripe fruit is scarce, fruits are often opened before they are completely ripe and the pulp is rapidly removed. In contrast, at the peak of fruit production, large quantities of uneaten fruits litter the bases of large saguaros. The pulp of fallen fruit is avidly consumed by mammals ranging in size from kangaroo rats *(Dipodomys merriami)* to coyotes *(Canis latrans)* and peccaries *(Pecari tajacu)*. At least one bird species, the mourning dove *(Zenaida macroura)*, consumes saguaro seeds from the ground without ingesting fruit pulp (Wolf et al. 2002). In some instances these doves seem to consume the "second harvest" of saguaro seeds defecated by mammalian seed dispersers (Haughey, unpublished data). Harvester ants *(Pogonomyrmex barbatus)* also remove numerous saguaro seeds (Steenbergh and Lowe 1977).

The large number of consumers of saguaro nectar, pollen, and fruit attests to the importance of this plant as a resource for desert animals. A rough estimation highlights the astounding amount of resources that saguaros provide annually. Assuming an average density of 100 reproductive plants per hectare, 300 flowers per plant, and 175 fruits per plant, the input of water and energy from saguaro floral nectar into the ecosystem is approximately 60 l of water and 27,500 kJ of energy per hectare. Pollen production represents an input of 203,750 kJ per hectare (Schmidt and Buchmann 1986). Saguaro fruits contribute 316 l of water and 1,375,000 kJ of energy per hectare (Wolf, unpublished). Stands with more than 100 reproductive individuals are not uncommon (Steenbergh and Lowe 1977; Wolf and Martínez del Rio 2003).

Although the identities of the animals that consume saguaro nectar, pollen, and fruit have been established (Alcorn et al. 1959; McGregor et al. 1962; Steenbergh and Lowe 1977), there is little quantitative data on the transfer of water and energy from saguaros into consumer populations (the Schmidt and Buchmann 1986 study on honeybees is an exception). The functional ecological importance (see Hurlbert 1997) of saguaros for Sonoran Desert ecosystems is presumably large, but it remains unmeasured.

White-Winged Doves

Western white-winged doves are the most frequent avian visitors to the saguaro's flowers and fruit (Haughey 1986). They migrate from southern Mexico (figure 7.1) into the Sonoran Desert to breed during the summer. The doves arrive in Arizona in early to mid-April, and breeding starts by early May (Neff 1940; Hensley 1959). Egg-laying commences in late May, and both males and females share in the incubation of the two eggs (Cottam and Trefethen 1968). Like other doves, nestlings are fed crop milk from hatching until they are about four days old, when the diet shifts to seeds from a variety of plants and saguaro fruit pulp (Wetmore 1920).

Why do white-winged doves migrate into the Sonoran Desert to breed during the most stressful portion of the year? The answer may be that the dove's breeding cycle is synchronized with the reproductive cycle of the saguaro (Shreve and Wiggins 1964; Haughey 1986), as we are currently documenting. Doves appear in the desert as the saguaros start blooming, they feed extensively at their flowers, and, when saguaro fruit is available, they eat it almost exclusively (Wolf and Martínez del Rio 2000). The ecological dependence of white-winged doves on saguaros (and other columnar cacti farther

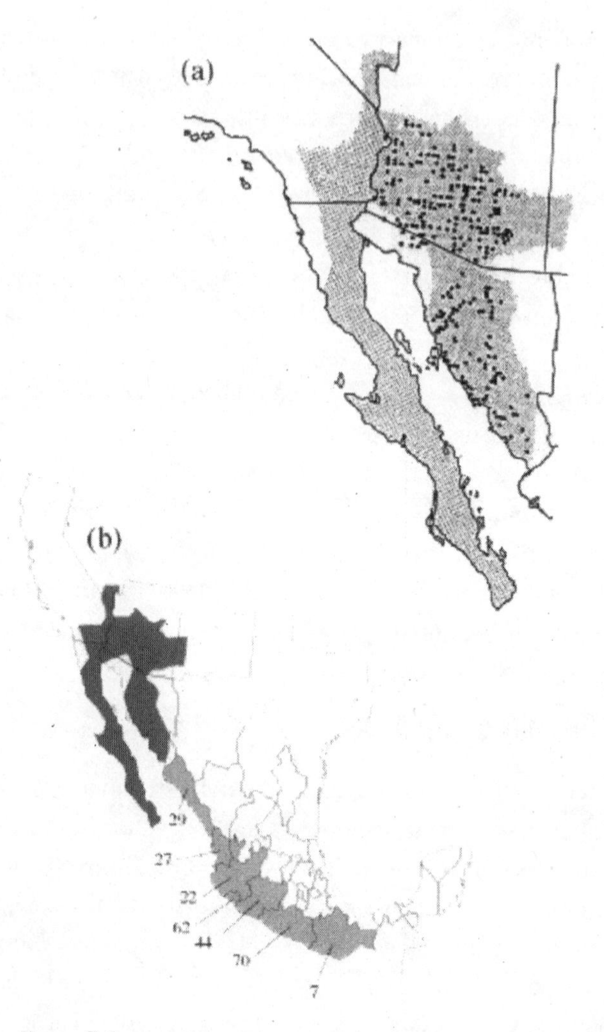

Figure 7.1. Breeding and wintering ranges of the western white-winged dove. The stippling in map (a) describes the breeding range; the dots in the eastern section of the range represent localities where saguaros are found (Turner et al. 1995). These doves also maintain a non-migratory population in Baja California (Saunders 1968). The range of western white-winged doves appears to be expanding both to the east, into New Mexico, and to the west, into California (Pruett et al. 2000). Map (b) shows the breeding (darker) and wintering ranges. Numbers represent the count of banded birds recovered in each of the west Mexican states in which the doves are known to winter.

south in the Sonoran Desert) is recognized in its Sonoran vernacular name, "paloma pitayera" (pigeon that eats columnar cactus fruit). Although the word "pitaya" is used in Sonora specifically for cacti in the genus *Stenocereus*, it is used generally for all columnar cacti fruit.

White-winged doves *(Zenaida asiatica)* are widespread in North and South America, with twelve morphologically distinct allopatric populations (Saunders 1968). Although there are no genetic data to determine the degree of isolation among all of these populations, the morphological differences among them have led some researchers to consider them subspecies (Saunders 1968, but see Browning 1990). Mitochondrial DNA evidence supports the contention that at least two populations are distinct (Pruett et al. 2000). These subspecies inhabit the United States and northern Mexico and have been characterized by Saunders (1968) as the western *(Z. a. mearnsii)* and eastern *(Z. a. asiatica)* white-winged dove. The recent range expansion of these two populations into eastern New Mexico and western Texas, respectively, may have created a zone of extensive intergradation between the two subspecies in the past 100 years (Pruett et al. 2000).

Of the twelve subspecies or populations defined by Saunders, only the two most northern subspecies *(Z. a. mearnsii* and *Z. a. asiatica)* are latitudinal migrants. To our knowledge, only the western white-winged dove relies extensively on saguaro nectar and fruit. The remaining eleven populations feed on a variety of seeds and agricultural grains (Cottam and Trefethen 1968). The reliance of western white-winged doves on saguaro may have led to morphological differences between this and other white-winged dove populations. The bills of western white-winged doves are significantly longer than expected from their size (figure 7.2). Females and males have culmen lengths that are 8 and 10 percent greater, respectively, than expected from a body size against culmen length regression based on other white-wing populations. We hypothesize that the longer bills of western white-winged doves facilitate probing efficiently into the deep flowers and fruits of saguaros.

The Dependence of Western White-Winged Doves on Saguaro

Anecdotal evidence of intensive use of saguaro flowers and fruits by western white-winged doves has been in the literature for years (reviewed

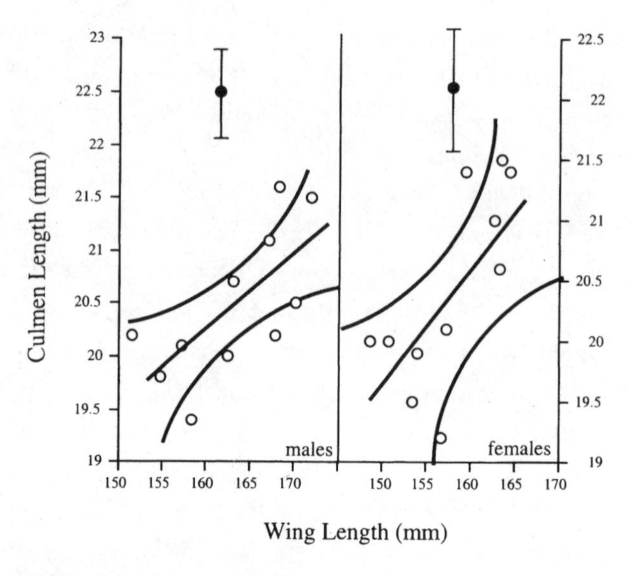

Figure 7.2. Bill lengths with respect to body size for western white-winged doves and other subspecies of *Zenaida asiatica*. Each point in this graph is the mean length of bill for each subspecies. Open circles represent all subspecies except western white-wings, for which the value is the filled circle. Error bars around this value are 95 percent confidence intervals.

by Haughey 1986 and Fleming et al. 1996). We used stable isotope analysis to determine the reliance of white-winged doves on saguaro quantitatively. Usually, the use by animals of different food types is estimated from pollen loads, fecal samples, analysis of crop contents, or foraging observations (Collins et al. 1990). Increasingly, however, the incorporation of the natural isotopic signatures of resources into a consumer's tissues is used to track use (Gannes et al. 1998). Stable isotopes are powerful tools for dietary analysis when applied to producer-consumer systems only if isotopically distinct food sources are available to consumers (Tieszen and Boutton 1989; Gannes et al. 1997). Saguaros exhibit such isotopically distinct signatures: the nutrients contained in saguaro contain a distinct carbon signature, and the water in its nectar and fruit has a unique hydrogen signature.

The isotopic signature of saguaro is the result of its crassulacean acid metabolism (CAM) photosynthetic pathway. Stem-succulent cacti, such as

TABLE 7.1. The Carbon Isotopic Composition of Saguaro

Nectar			Fruit		
5/24/98	6/11/98	6/29/98	6/20/98	7/20/98	8/3/98
−13.0 ± 0.3 (8)	−12.9 ± 0.2 (11)	−12.3 ± 0.1 (9)	−12.9 ± 0.1 (10)	−13.0 ± 0.2 (9)	−13.4 ± 0.2 (10)

Note: The $\delta^{13}C$ (‰) VPDB (Vienna Pee Dee Belemnite) of saguaro resources as a function of date sampled varies little and is enriched in ^{13}C. Means ± SE and sample sizes (*n*) are presented.

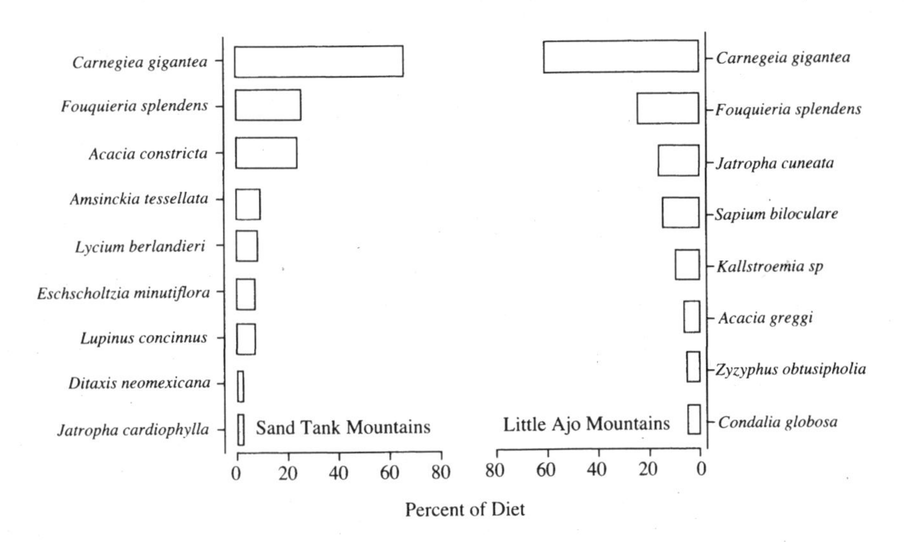

Figure 7.3. Composition of the diet of western white-winged doves at the Sand Tank and the Little Ajo Mountains in Arizona.

the saguaro, tend to exhibit obligate-CAM photosynthesis and thus have isotopic signatures similar to those of C4 species that are relatively enriched in ^{13}C (table 7.1; Bender 1971; Ehleringer 1989). The alternative food sources of saguaro contain a different isotopic signature. An extensive survey of white-winged dove crop contents revealed that the seeds from eight plant species other than saguaro appeared frequently in the dove's diet (figure 7.3; see also Haughey 1986). The saguaro and these C3 plant species accounted for more than 90 percent of the white-wing's crop contents. These eight species all have a C3 photosynthetic pathway, and their average carbon isotopic composition is $\delta^{13}C = -24.9 \pm 0.3$ SE ‰. Figure 7.4 describes the changes in isotopic composition of dove tissues throughout the dove's stay at a site in

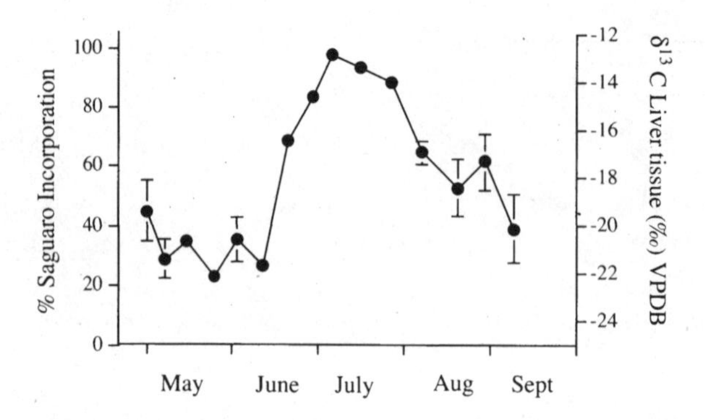

Figure 7.4. Changes in the isotopic composition of western white-winged dove diets through their stay in the Sonoran Desert.

southern Maricopa County, Arizona. Saguaro is not only the most frequent item in the dove's diet but is also the primary source of incorporated carbon for a large fraction of the breeding season. In July, during the peak of saguaro use, the isotopic composition of the doves' tissues was almost indistinguishable from that of saguaro. In isotopic terms, breeding white-winged doves are warm, feathered fragments of saguaro flying about in the desert.

Two isotopically distinct sources of water are available for desert-dwelling white-winged doves: the nectar and fruit of saguaro, and surface water in man-made tanks and natural tinajas (natural water catchments). We have found that water from the nectar and fruit of the saguaro is greatly enriched in deuterium, a heavy isotope of hydrogen, relative to surface water (table 7.2). The body water of white-winged doves becomes deuterium-enriched with increased incorporation of saguaro carbon (figure 7.5). White-winged doves derive not only nutrients but also water from saguaro.

What Do Saguaros Get from Doves?

Doves obtain nutrients and water from saguaros. What do saguaros receive in return? Because doves visit both the flowers and the fruit of saguaros, they are potential pollinators and seed dispersers. When saguaros are in bloom, white-winged doves visit their flowers frequently and appear to carry large pollen loads (figure 7.6). At a coastal site in Sonora, Fleming et al. (1996) found that these birds visited saguaro flowers an order of magnitude

TABLE 7.2. The Hydrogen Isotopic Composition of Saguaro Water

	Nectar		Fruit		
	5/24/98	6/11/98	6/20/98	7/20/98	8/20/98
Saguaro	19.6 ± 3.3 (9)	26.4 ± 3.2 (13)	47 ± 3.1 (17)	50.5 ± 2.3 (47)	41.3 ± 3.5 (19)
Surface Water	−22.4	−20.2	−18.8	−34.1	−36.2

Note: At all dates sampled, the hydrogen isotopic composition of saguaro water, estimated by δD (‰) VSMOW (Vienna Standard Mean Ocean Water), is very different from other sources of water in the desert. Means ± SE, and sample sizes (*n*) are presented. Note that surface water enriches in δD slightly as a result of evaporative fractionation before the onset of monsoon rains. After rains, the composition of surface water approaches that of rainfall. The temporal variation in isotopic composition in both saguaro resources and rainwater is small relative to the turnover of the deuterium pool in doves. However, the use of deuterium as a resource tracer for saguaro consumption requires monitoring the composition of both sources.

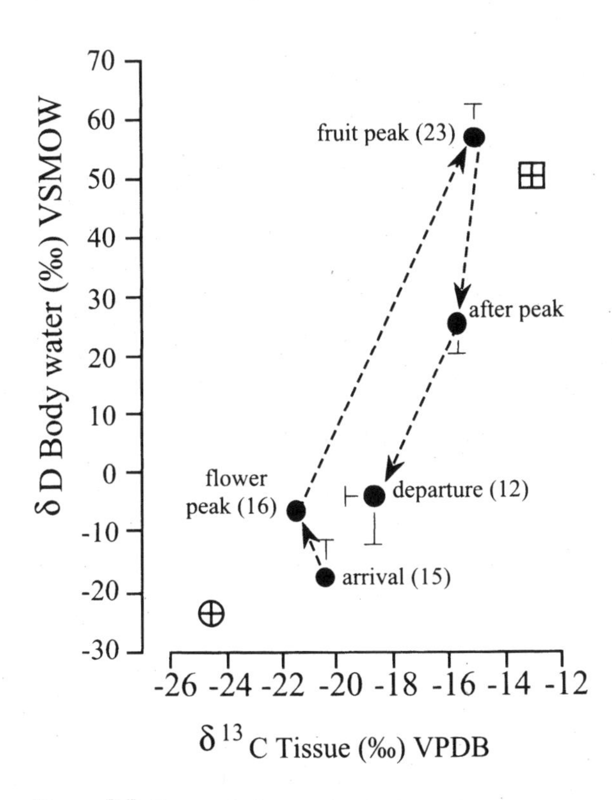

Figure 7.5. Seasonal changes in the isotopic composition of carbon and deuterium in the tissues and body water of white-winged doves.

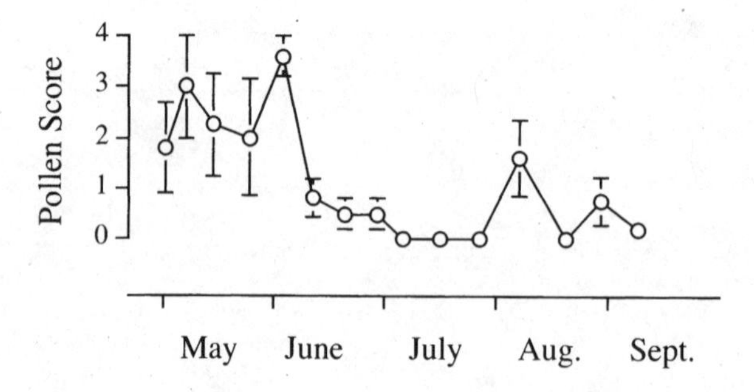

Figure 7.6. Qualitative estimate of the saguaro pollen load carried by each dove, obtained by swabbing the bill, crown, cheeks, and chin of each bird with a clean strip of transparent adhesive tape.

more frequently than bats. Clearly, white-winged doves are important pollinators of saguaro. Curiously, however, at this site the most frequent visitors, and presumably pollinators, of saguaro flowers were introduced honeybees. Honeybees visited saguaro flowers three times more frequently than doves. It is unknown if the patterns found by Fleming et al. can be generalized to all areas of the Sonoran Desert. Honeybee population densities in the desert can vary enormously in space and time (Schmidt and Edwards 1998), and it is likely that their importance as saguaro pollinators, relative to that of doves, varies as well.

White-winged doves are saguaro pollinators. Are they also saguaro seed dispersers? They have powerful gizzards (Goodwin 1983), and they destroy the vast majority of the saguaro seeds that they ingest (reviewed by Sosa 1997). However, they regurgitate saguaro fruit pulp to their squabs (Neff 1940; Wolf and Martínez del Rio 2002), and, in the messy process, some seeds fall intact beneath their nests (Olin et al. 1989). Desert-dwelling white-winged doves nest in large trees such as mesquites *(Prosopis velutina)*, paloverdes (*Cercidium* spp.), and ironwoods *(Olneya tessota)*, which can act as nursery plants for saguaro seedlings (McAuliffe 1984). Therefore, Olin et al. (1989) and Sosa (1997) have hypothesized that doves may play an important dispersal role. However, relative to the number of seeds destroyed by doves, the number of seeds dropped under nests is minute. Olin et al. reported that the number of seeds spilled under nests ranged from 57 to 675 per season. A single dove ingests approximately 3,400 g of saguaro fruit pulp, containing about 280,000 seeds, per season (Wolf and Martínez del Rio 2000). Although white-winged doves are the most important consumers of saguaro fruit pulp, they are, paradoxically, better characterized as seed predators than as seed dispersers. The bulk of effective dispersal of saguaro seeds is probably performed by birds with gentler guts, such as Gila woodpeckers, verdins, ash-throated flycatchers, curve-billed thrashers, and cactus wrens (Sosa 1997).

In summary, saguaros and white-winged doves maintain a lopsided interaction. Doves benefit from saguaros; they obtain both nectar and nutritious fruit pulp from them. We suspect that saguaros are crucial for the productivity and persistence of white-winged dove populations in the Sonoran Desert. All data suggest that desert-dwelling white-winged doves are saguaro specialists. However, despite the apparent specialization of doves on saguaros, the effect on these cacti is not one of unquestionable benefit. Although saguaros receive significant pollination services from doves, the birds also destroy an enormous number of saguaro seeds. Without more data on the relative importance of doves as pollinators, it is difficult to assess the cost-benefit balance of their effect on saguaros. The importance of white-winged doves as saguaro pollinators may increase if, as expected, tracheal and varroa mite infestations lead to severe declines in feral honeybee populations in the Sonoran Desert (Watanabe 1994; Buchmann and Nabhan 1996; Loper 1997; Kearns et al. 1998).

Playing God with the White-Winged Dove

The fate of white-winged doves is not only linked to that of saguaros but is also dependent on the vagaries of the relationship of humans with the land. The dramatic population fluctuations experienced by white-winged dove populations in Arizona in the past 150 years can be largely explained by changes in how humans have perceived doves (as pests, quarry, and a fragile resource) and by how humans have used the land that they share with doves and saguaros (Alcock 1993). This review of the interaction of white-winged doves with humans in Arizona relies heavily on the excellent summary provided by Brown (1989) with a minor, but significant, difference: we suggest that our perception of population trends of white-winged doves in Arizona may be biased by emphasis on the large and very variable populations that nested colonially in extensive riparian mesquite bosques and that presumably fed on agricultural products. These populations provide the bulk of birds taken by Arizona hunters and thus are the ones that have been most intensely studied (Cunningham et al. 1997). The populations of desert-dwelling, saguaro-dependent doves have more scattered nests (Arnold 1943; Wigal 1973), are more difficult to hunt (Martínez del Rio and Wolf, pers. obs.), and have received significantly less study. Their population trends in the past century are not well understood.

The size of white-winged dove populations prior to large-scale agriculture in Arizona is unknown. They are reported to have been common, even numerous (Bendire 1892), but their numbers seemed to have been lower than in the late 1800s, when increased cereal production appears to have led to a large population increase (Cottam and Trefethan 1968). Like other members of their genus (such as eared doves, *Zenaida auriculata*), white-winged doves are facultative colonial nesters (Bucher 1992). It is unknown whether or not they nested colonially before large-scale agriculture. However, by the beginning of the twentieth century, several large colonies were well established in the then-extensive riparian mesquite bosques along the Santa Cruz and Gila Rivers. Before 1941 they were managed as de facto agricultural pests (Wetmore 1920). Hunting seasons overlapped with breeding, and bags were generous (thirty-five birds per day; Brown 1989). As a result of overhunting and the destruction of nesting habitat, and perhaps because of the conversion of grain fields to cotton (O'Connor 1939), white-winged dove populations collapsed (Neff 1940). After the hunting season opening date was moved from August 1 (when doves are still nesting) to September 1, populations increased again until the 1960s (Cottam and Trefethan 1968).

In the 1960s, white-winged doves were so abundant that the bag was again set at twenty-five birds. By 1968, the population began to decline again, apparently as a result of the combined effect of nesting habitat loss and a dramatic reduction in cereal production, and possibly from overharvest (Smith 1983). From 1967 to 1980, the number of birds killed by hunters dropped from 700,000 to fewer than 100,000 (Brown 1989). Recent data on white-winged dove populations are unavailable, but the numbers killed by hunters remain low (D. E. Brown, pers. comm.). The morning feeding flights of white-winged doves that were described by Cottam and Trefethen (1968: 225) as "one of the great natural wonders of Arizona" may have been artificial. These flights are, as a consequence of human activities, a thing of the past. In the words of Alcock (1993: 47), "we have played god with the white-winged dove."

We suspect that the dramatic fluctuations described above accurately illustrate the close association of white-winged doves with agricultural fields and riparian thickets. It is not certain that desert-dwelling, saguaro-dependent white-winged doves followed the same trends. The foraging habits, social behavior, and demography of desert-dwelling doves appear to differ from those of birds dependent on agricultural products. The enormous population fluctuations exhibited by white-winged doves can only be explained by a very productive population that depended strongly on clumped abundant resources and that had large tracts of dense nesting habitat (Bucher 1982). These are not characteristics that typify desert-nesting doves. We suspect, but have little data to support this suspicion, that the populations of desert-dwelling and saguaro-dependent doves have fluctuated much less than those of birds that nest colonially in river bottom thickets and that rely on cereals.

During the breeding season, more than 60 percent of the diet of desert-dwelling white-winged doves is saguaro (Haughey 1986; Wolf and Martínez del Rio 2000). Although saguaro groves can produce large amounts of fruit, these numbers cannot be compared with the enormous productivity of irrigated cereal fields (Smith 1983). Saguaro represents only 2–5 percent of the diet of the last white-winged dove colonies that still depend on agricultural products in Arizona (Cunningham et al. 1997). The bulk of the diet of these doves consists of barley, milosorghum, safflower, and corn. Nesting pairs of desert-dwelling doves are scattered, rather than clumped in colonies (Arnold 1943; Haughey, Wolf, and Martínez del Rio, unpublished data). Birds feeding in agricultural fields are noticeably gregarious (Butler 1977; Cunningham et al. 1997), whereas desert doves feeding on saguaro feed singly or, more rarely, in pairs (Wolf and Martínez del Rio 2002). Finally, the number of clutches per season, and hence the productivity, of white-winged doves nesting in the desert is lower (one clutch per season) than that of colonially nesting birds that feed on grain (sometimes two clutches per season; Brown 1977).

In summary, nesting habitat availability and food distribution and abundance seem to have a profound influence on the biology of white-winged doves. South American eared doves, *Z. auriculata*, show similar flexibility in reproductive tactics and feeding behavior. When food and nesting sites are dispersed, they feed and nest solitarily. When food is concentrated and interdispersed with extensive suitable nesting habitat, they feed gregariously and nest in large, productive colonies (Murton et al. 1974; Bucher 1982). White-winged doves in the Sonoran Desert may be unique in that, at least during this century, a colonially nesting population that fed on cultivated grain co-existed with a solitary nesting population that fed primarily on the fruit and nectar of a single plant species.

Future Research Directions

Several aspects of the interaction between saguaros, doves, and other pollinators and seed dispersers require further investigation. Here we identify and develop three questions that we believe merit research.

1. What is the effect of bees on the pollination ecology of saguaro in general, and on the foraging behavior of white-winged doves in particular? In a well-remembered event, Seri elders date the collection of the first honeycomb in the Sonora coastal area at around 1900 (Felger and Moser 1991). Given the relatively recent introduction of honeybees into the Sonoran Desert, it is of interest to determine their impact on saguaro pollination and on the interaction of saguaros with white-winged doves, the native saguaro pollinators. Herrera (2000) has demonstrated that the composition of the pollinator assemblage of a plant can influence outcrossing, the viability of seeds, and the establishment of seedlings. Honeybees probably have a dual impact on saguaro pollination. First, because honeybees have minuscule foraging ranges relative to those of doves (McGregor et al. 1959), their pollen shadows are probably smaller (Hamrick 1987). Second, honeybees are efficient nectar foragers, and they are capable of rapidly depleting the nectar and pollen available for other pollinators (Schmidt and Buchmann 1986; Wolf and Martínez del Rio 2002), such as doves, thus reducing their per-flower visitation rates (Wills et al. 1990; Paton 1993). The introduction of honeybees into the Sonoran Desert may have influenced the reproductive biology of saguaros by reducing pollinator neighborhoods and by altering the foraging patterns of native pollinators.

There is little evidence to evaluate the effect of feral and domestic bees on the reproductive biology of saguaros and on their interaction with native pollinators. Gathering these data presents a unique opportunity for understanding the role of an introduced invasive species on a keystone resource provider and on its interaction with native pollinators. The effect of honeybees on saguaro pollination can be studied using both comparative and experimental approaches: a comparative study of saguaro pollination at sites with and without honeybees may yield interesting insights. The experimental removal of honeybees from large areas is difficult but feasible (G. M. Loper, pers. comm.). The almost worldwide introduction, and subsequent naturalization, of honeybees is a large unintended experiment with probably important but generally undocumented consequences. Few places in the world are free of feral or managed bees (Buchmann and Nabhan 1996). The long-term response of a community of pollinators to the removal of honeybees would be exceedingly instructive. We predict increased visitation by doves—and other avian pollinators—to saguaro flowers in response to honeybee removals (or at sites without honeybees, if these can be found). We also predict increased outcrossing due to pollination in saguaro resulting from the increased importance of long-ranging pollinators that produce longer seed shadows.

2. Saguaros have been characterized as a keystone species. How important are saguaros as keystone resource providers in Sonoran Desert ecosystems? The term "keystone" is controversial (see Mills et al. 1993 and Power et al. 1996). Hurlbert (1997) has argued that keystone has come to mean little more than that a species is important for something. As one alternative to the keystone species notion, he proposes the concept of general functional importance. For a producer, such as the saguaro, functional importance can be defined as the sum, over all consumers, of the changes in productivity that would occur on removal of the consumer from the biocenosis. Although this concept cannot be easily determined empirically, it suggests measurements that allow assessing the significance of a species for a defined subset of consumers—in this case, the animals that eat saguaro nectar, pollen, and fruit pulp. An essential ingredient in the assessment of the functional importance of a producer is its contribution to the energy, nutrient, and water budgets of its consumers (Reiners 1986).

Excluding an unethical experiment in which saguaros are removed from a site, how can one evaluate the functional importance of saguaros? At the very least, how can the flow of energy and materials from saguaros into consumers be estimated? One can use the saguaro's distinctive isotopic carbon and deuterium signature to track the incorporation of saguaro carbon and water into consumers' tissues and body water. Wolf and Martínez del Rio (2000) outlined a method that combines isotopic measurements with physiological measurements to estimate the flow of carbon or energy and water from saguaro into consumer populations. The combination of field measurements of water and energy fluxes in consumers with isotopic estimates of resource use is ideally suited to estimate the functional importance of saguaros. Saguaro populations are affected by human-mediated activities such as grazing (Abouhaidar 1992; Parker 1993), the expansion of exotics (Pierson and Turner 1998), and urban development (Niering and Whittaker 1965). Determining the functional importance of saguaros will allow assessing the indirect effects that these threats on saguaro populations can have on other members of Sonoran Desert biotic communities.

3. What is the status of saguaro-dependent white-winged dove populations? White-winged doves are both migrant pollinators and a game species that is avidly hunted. Thus, their conservation and management pose peculiar challenges. As mentioned before, we have little information on population trends for desert-dwelling white-winged doves. Given their game status and their importance as saguaro pollinators, a program of population monitoring in Arizona is needed. Because white-wings are migratory, their populations depend on conditions at both the breeding and wintering areas. Their status must be determined, and their management must span both Arizona and Mexico.

An extensive banding program conducted by the Arizona Game and Fish Department provided detailed information on the migration of Arizona's white-winged doves in the years before the population collapsed (that is, before 1970; Stair 1970). The results of this study revealed that most of the birds banded in Arizona spend the winter in the Pacific coastal plains and foothills from southern Sinaloa to Guerrero and Oaxaca (see figure 7.1). Of the 265 birds banded in Arizona and recovered dead in Mexico, only 4 were killed in the state of Sonora, suggesting that birds fly more or less directly between the breeding and wintering grounds (Cottam and Trefethen 1968: table 3). The most widespread native vegetation types in the areas where bands were recovered are deciduous and subdeciduous tropical woodlands (Rzedowsky 1975). However, these areas also contain expanding pockets of agriculture and secondary vegetation and small areas with thornscrub and, at higher elevations, pine-oak forests (Rzedowsky 1975). Most of the birds banded in Arizona were "agricultural" birds that were recovered by Mexican hunters in agricultural cultivated areas (Kufeld 1963).

Thus, the habitat preferences of desert-breeding white-winged doves and the status of their habitat are unclear. Native forests in western Mexico are being rapidly cleared into agricultural fields and pasturelands (Ceballos and García 1995), and white-winged doves are hunted by both subsistence and sport hunters throughout the country (Oficina General de Fauna Silvestre 1968). Information on the habitats and resources used by white-winged doves in the winter, as well as data on numbers of birds harvested by Mexican hunters, can help to interpret their population trends in the breeding grounds and can provide guidelines for their management in both Mexico and Arizona.

ACKNOWLEDGMENTS

We thank Gary Nabhan and other migrant pollinator collaborators for having invited us to participate in the migrant pollinator initiative. The chapter benefited from the constructive comments of Bill Calder, Richard Felger, Ted Fleming, and Gary Nabhan. The research of Wolf and Martínez del Rio described here was funded by the Turner Endangered Species Fund Foundation, the National Science Foundation, the Ecological Society of America, and the University of Arizona.

LITERATURE CITED

Abouhaidar, F. 1992. Influence of livestock grazing on saguaro seedling establishment. Pp. 57–61 in C. P. Stone and E. S. Bellantoni, eds., Proceedings of the Symposium on Research in Saguaro National Monument, Jan. 23–24, 1991. Southwest Parks and Monuments Association, Globe, Ariz.

Alcock, J. 1993. The Masked Bobwhite Rides Again. University of Arizona Press, Tucson.

Alcorn, S. M., S. E. McGregor, and G. Olin. 1959. Pollination of the saguaro cactus by doves, nectar-feeding bats, and honey bees. Science 133:1594–95.

Arnold, L. R. 1943. A study of the factors influencing the management of and a suggested management plan for the western white-winged dove in Arizona. Pittman-Robertson Proj. 9R, Arizona Game and Fish Department, Phoenix.

Bender, M. M. 1971. Variations in the $^{13}C/^{12}C$ ratios of plants in relation to the pathway of photosynthetic carbon dioxide fixation. Phytochemistry 10:1239–44.

Bendire, C. 1892. Life histories of North American birds with special reference to their breeding habits and eggs, with twelve lithographic plates. Special Bulletin U.S. National Museum 1:i–446.

Bowers, J. E. 1996. Environmental determinants of flowering date in the columnar cactus *Carnegiea gigantea* in the northern Sonoran Desert. Madroño 43:68–84.

Brown, D. E. 1977. White-winged doves. Pp. 247–72 in G. C. Sandesron, ed., Management of Migratory Shore and Upland Game Birds in North America. University of Nebraska Press, Lincoln.

———. 1989. Arizona Game Birds. University of Arizona Press, Tucson.

Browning, M. R. 1990. Taxa of North American birds described from 1957 to 1987. Proceedings of the Biological Society of Washington 103:432–51.

Bucher, E. H. 1982. Colonial breeding of the eared dove *(Zenaida auriculata)* in northeastern Brazil. Biotropica 14:255–61.

———. 1992. The causes of extinction of the passenger pigeon. Current Ornithology 9:1–36.

Buchmann, S. M., and G. P. Nabhan. 1996. The Forgotten Pollinators. Island Press, Washington, D.C.

Butler, W. I. 1977. A white-winged dove nesting study in three riparian communities on the lower Colorado River. Master's thesis, Arizona State University, Tempe.

Ceballos, G., and A. García. 1995. Conserving Neotropical biodiversity: The role of dry forests in western Mexico. Conservation Biology 9:1349–53.

Collins, B. G., G. J. Grey, and S. McNee. 1990. Foraging and nectar use in nectarivorous bird communities. Studies in Avian Biology 13:110–22.

Cottam, C., and J. B. Trefethen. 1968. Whitewings: The Life History, Status, and Management of the White-winged Dove. D. Van Nostrand, Princeton, N.J.

Cunningham, S. C., R. W. Engel-Wilson, P. M. Smith, and W. B. Ballard. 1997. Food habits and nesting characteristics of sympatric mourning doves in Buckeye–Arlington Valley, Arizona. Technical Report 26, Arizona Game and Fish Department, Phoenix.

Ehleringer, J. R. 1989. Carbon isotope ratios and physiological processes in aridland plants. Pp. 41–54 in P. W. Rundel, J. R. Ehleringer, and K. A. Nagy, eds., Stable Isotopes in Ecological Research. Springer-Verlag, New York.

Felger, R. S., and M. B. Moser. 1991. People of the Desert and the Sea: Ethnobotany of the Seri Indians. University of Arizona Press, Tucson.

Fleming, T. H., M. D. Tuttle, and M. A. Horner. 1996. Pollination biology and the relative importance of nocturnal and diurnal pollinators in three species of Sonoran Desert columnar cacti. Southwestern Naturalist 41:257–69.

Gannes, L. Z., C. Martinez del Rio, and P. Koch. 1998. Natural abundance variations in stable isotopes and their uses in animal physiological ecology. Comparative Biochemistry and Physiology 119A:725–37.

Gannes, L. Z., D. M. O'Brien, and C. Martínez del Rio. 1997. Stable isotopes in animal ecology: Assumptions, caveats and a call for more laboratory experiments. Ecology 78:1271–76.

Goodwin, D. 1983. Pigeons and Doves of the World. Cornell University Press, Ithaca, N.Y.

Greenhouse, R. 1979. The Iron and Calcium Content of Some Traditional Pima foods and the Effects of Preparation Methods. Arizona State University, Tempe.

Hamrick, J. L. 1987. Gene flow and distribution of genetic variation in plant populations.

Pp. 53–67 in K. M. Urbanska, ed., Differentiation Patterns in Higher Plants. Academic Press, London.
Haughey, R. A. 1986. Diet of desert-nesting western white-winged doves, *Zenaidia asiatica mearnsii*. Master's thesis, Arizona State University, Tempe.
Hensley, M. M. 1959. Notes on the nesting of selected species of birds of the Sonoran Desert. The Wilson Bulletin 71(1):86–92.
Herrera, C. M. 2000. Flower-to-seedling consequences of different pollination regimes in an insect-pollinated shrub. Ecology 81:15–29.
Hurlbert, S. 1997. Functional importance vs. keystoneness: Reformulating some questions in theoretical ecology. Australian Journal of Ecology 22:369–82.
Kearns, C. A., D. W. Inouye, and N. M. Waser. 1998. Endangered mutualisms: The conservation of plant pollinator interactions. Annual Review of Ecol. Syst. 29:83–112.
Kufeld, R. C. 1963. Summary and analysis of data for mourning and white-winged doves banded in Arizona. Special Report, Arizona Game and Fish Department, Phoenix.
Loper, G. M. 1997. Over-winter losses of feral honeybee colonies in southern Arizona, 1992–1997. American Bee Journal 35:446.
McAuliffe, J. R. 1984. Sahuaro–nurse tree associations in the Sonoran Desert: Competitive effects of sahuaros. Oecologia 64:319–21.
McGregor, S. E., S. M. Alcorn, E. B. Kurtz, and G. D. Butler. 1959. Bee visitors to saguaro flowers. Journal of Economical Entomology 52:1002–4.
McGregor, S. E., S. M. Alcorn, and G. Olin. 1962. Pollination and pollinating agents of the saguaro. Ecology 43:259–67.
Mills, L. S., M. E. Soule, and D. F. Soak. 1993. The keystone species concept in ecology and conservation. Bioscience 43(4):219–44.
Murton, R. K., E. H. Bucher, M. Nores, and J. Reartes. 1974. The ecology of the eared dove *(Zenaida auriculata)* in Argentina. The Condor 76:80–88.
Neff, J. A. 1940. Notes on nesting and other habits of the western white-winged dove in Arizona. Journal of Wildlife Management 4:279–90.
Niering, W. A., and R. H. Whittaker. 1965. The saguaro problem and grazing in southwestern national monuments. National Parks Magazine 39:4–9.
O'Connor, J. 1939. Game in the Desert. Derrydale Press, New York.
Oficina General de Fauna Silvestre. 1968. La paloma de alas blancas en México. Subsecretaría de Recursos Forestales y Fauna Silvestre, Distrito Federal, México.
Olin, G., S. M. Alcorn, and J. M. Alcorn. 1989. Dispersal of viable saguaro seeds by white-winged doves *(Zenaida asiatica)*. The Southwestern Naturalist 34:281–84.
Parker, K. C. 1993. Climatic effects on regeneration trends for two columnar cacti in the northern Sonoran Desert. Annals of the Association of American Geographers 83:452–74.
Paton, D. C. 1993. Honeybees in the Australian environment. BioScience 43:95–103.
Peebles, R. H., and H. Parker. 1946. Watching the saguaro bloom. Desert Plant Life 18:55–60.
Pierson, E. A., and R. A. Turner. 1998. An 85-year study of saguaro *(Carnegiea gigantea)* demography. Ecology 79:2676–93.
Power, M. E., D. Tilman, J. Estes, B. A. Menge, W. J. Bond, L. S. Mills, G. Daily, J. C. Castilla, J. Lubchenco, and R. T. Paine. 1996. Challenges in the quest for keystones. BioScience 46:609–20.
Pruett, C. L., S. E. Kenke, S. M. Tanksley, M. F. Small, K. M. Hogan, and J. Roberson. 2000.

Mitochondrial DNA and morphological variation of white-winged doves in Texas. The Condor 102:871–80.

Reiners, W. A. 1986. Complementary models for ecosystems. American Naturalist 127:59–73.

Ross, W. 1944. The present-day dietary habits of the Papago Indians. Master's thesis, University of Arizona, Tucson.

Rzedowsky, J. 1975. Vegetación de México. Limusa, Distrito Federal, México.

Saunders, G. B. 1968. Seven new white-winged doves from Mexico, Central America, and the southwestern United States. North American Fauna 65, U.S. Bureau of Sports Fisheries and Wildlife, Washington, D.C.

Schmidt, J. O., and S. L. Buchmann. 1986. Floral biology of the saguaro, *Cereus giganteus*. Oecologia 69:491–98.

Schmidt, J. O., and J. F. Edwards. 1998. Ecology of feral and africanized honeybees in Organ Pipe Cactus National Monument. First Conference on Research and Resource Management in Southern Arizona National Park Areas, Tucson.

Shreve, F. 1945. The saguaro, cactus camel of Arizona. National Geographic Magazine 88:69–74.

———. 1951. Vegetation of the Sonoran Desert. Carnegie Institution of Washington Publication 591:192.

Shreve, F., and I. L. Wiggins. 1964. Vegetation and Flora of the Sonoran Desert. Stanford University Press, Stanford, Calif.

Smith, P. 1983. Where are all the whitewings? Arizona Wildlife Views 26:6.

Sosa, V. 1997. Dispersal and recruitment ecology of columnar cacti in the Sonoran Desert. Ph.D. diss., University of Miami, Fla.

Stair, J. 1970. Chronology of the nesting season of white-winged doves *Zenaida asiatica mearnsii* (Ridgway) in Arizona. Master's thesis, University of Arizona, Tucson.

Steenbergh, W. F., and C. H. Lowe. 1977. Ecology of the Saguaro II: Reproduction, Germination, Establishment, Growth, and Survival of the Young Plant. U.S. Government Printing Office, Washington, D.C.

Thackery, F. A., and A. R. Leding. 1929. The giant cactus of Arizona: The use of its fruit and other cactus fruits by the Indians. The Journal of Heredity 20:400–414.

Tieszen, L. L., and T. W. Boutton. 1989. Stable carbon isotopes in terrestrial ecosystem research. Pp. 167–95 in P. W. Rundel, J. R. Ehleringer, and K. A. Nagy, eds., Stable Isotopes in Ecological Research. Springer-Verlag, New York.

Turner, R. M., J. M. Bowers, and T. L. Burgess. 1995. Sonoran Desert Plants, an Ecological Atlas. University of Arizona Press, Tucson.

Watanabe, M. E. 1994. Pollination worries rise as honeybees decline. Science 265:1170.

Weidner-Zoehfeld, K., and P. Mirocha. 1997. Cactus Cafe: A Story of the Sonoran Desert. Soundprints, New York.

Wetmore, A. 1920. Observations on the habits of the white-winged dove. Condor 22:140–46.

Wigal, D. D. 1973. A survey of the nesting habitats of the white-winged dove in Arizona. Special Report No. 2, Arizona Game and Fish Department, Phoenix.

Wills, R. T., M. N. Lyons, and D. T. Bell. 1990. The European honeybee in western Australian Kwongan: Foraging preferences and implications for management. Proceedings of the Ecological Society of Australia 16:167–76.

Wolf, B. O., J. Babson, and C. Martínez del Rio. 2002. Saguaros, doves, and isotopes: Differential water and carbon acquisition by animals feeding on a single resource. Ecology 83(3):1286–93.

Wolf, B. O., and C. Martínez del Rio. 2000. Use of saguaro fruit by white-winged doves: Isotopic evidence of a tight ecological association. Oecologia 124:536–43.

———. 2002. Flight path: Saguaro-feeding by desert birds. Birding (August):324–25.

———. 2003. How important are CAM succulents as sources of water and nutrients for desert consumers? A review. Isotopes in Environment and Health Science 39:53–67.

CHAPTER 8

The Interchange of Migratory Monarchs between Mexico and the Western United States, and the Importance of Floral Corridors to the Fall and Spring Migrations

LINCOLN P. BROWER AND ROBERT M. PYLE

Most conservation activities focus on the diminution of species diversity while largely ignoring a recent theme of equal importance: endangered biological phenomena. These are particularly notable aspects of the life histories of animal or plant species involving very large numbers of individuals—the species per se need not be in peril; rather, some spectacular phenomenon it exhibits is at stake. This concept was proposed by Brower and Pyle (1980), was formalized in Wells et al. (1983: xxxii) as "threatened phenomenon" and in Pyle (1983a, 1983b, 1983c), and was elaborated in Brower and Malcolm (1989, 1991) and Malcolm (1993). Examples of endangered phenomena include scores of current animal migrations that are being disrupted by accelerating habitat modification throughout the world (see also Brower 1997).

Monarch butterfly migration in the United States, Canada, and Mexico, together with the extraordinary aggregations of up to tens of millions of individuals overwintering in Mexico and hundreds of thousands in Alta California, well exemplify the concept of endangered biological phenomena. The monarch (*Danaus plexippus* [L.], Nymphalidae) is a member of the tropical subfamily Danainae that contains 157 species known collectively as "the milkweed butterflies" because the caterpillars feed on plants that produce milky latex, particularly the Asclepiadaceae (Ackery and Vane-Wright 1984, 1985). Several danaines are known to migrate, but the monarch is unique in having a very long-distance and bird-like migration that involves up to four generations per annual cycle (Brower 1995a). This complex behavior evolved in response to, and allows the monarch to exploit many, if not all, of the 106 species of *Asclepias* across the North American continent, where it has become one of the most abundant and conspicuous butterflies in the world. Remarkably, and in contrast to vertebrate migrations, the monarch's orientation and navigation to its overwintering sites is carried out by descendants three or more generations removed from their migrant forebears. Its fall migration, therefore, is an inherited behavior pattern with no opportunity to learn the migratory routes (Brower 1996). Prior understanding of the routes and corridors of monarch migration in North America is rapidly evolving due to new information, especially in the West.

Previous Models of Western Monarch Movements

Based on an amalgam of ideas available at the time, including recoveries of monarchs tagged and released during the fall migration summarized by Urquhart (1960), Zahl (1963) published a map suggesting that monarchs migrate southward from both the western and the eastern United States to central Mexico. However, the absence of evidence over the succeeding years on the presumed overwintering area in Mexico led to the general acceptance of a model involving two separate migratory populations in North America (see Brower 1995a: figs. 1a, 1b). According to this scenario, one population breeds during the summer east of the Rocky Mountains and migrates to and overwinters on twelve mountain massifs in central Mexico (Brower et al. 2002), while a much smaller population breeds west of the Rocky Mountains and migrates to and overwinters in numerous sites principally along the Pacific Coast in California.

Even though Brower (1995a) pointed out that the demarcation between the two is largely hypothetical and the degree of natural interchange between them is largely unknown, the model of two distinct populations has become almost universally accepted and has been depicted on all subsequent maps of the migration, including one in a recent federal survey of monarch conservation efforts (State and Federal Monarch Activities 1997) and another in the large-circulation *Sunset* magazine (Jenson 2000) that perpetuated the idea with the caption "The Great Divide." Thus the East-West bifurcation of monarchs on either side of the Continental Divide has become a virtual canon in American natural history. This model is represented by figure 8.1.

Several investigators have indicated doubts about this traditional

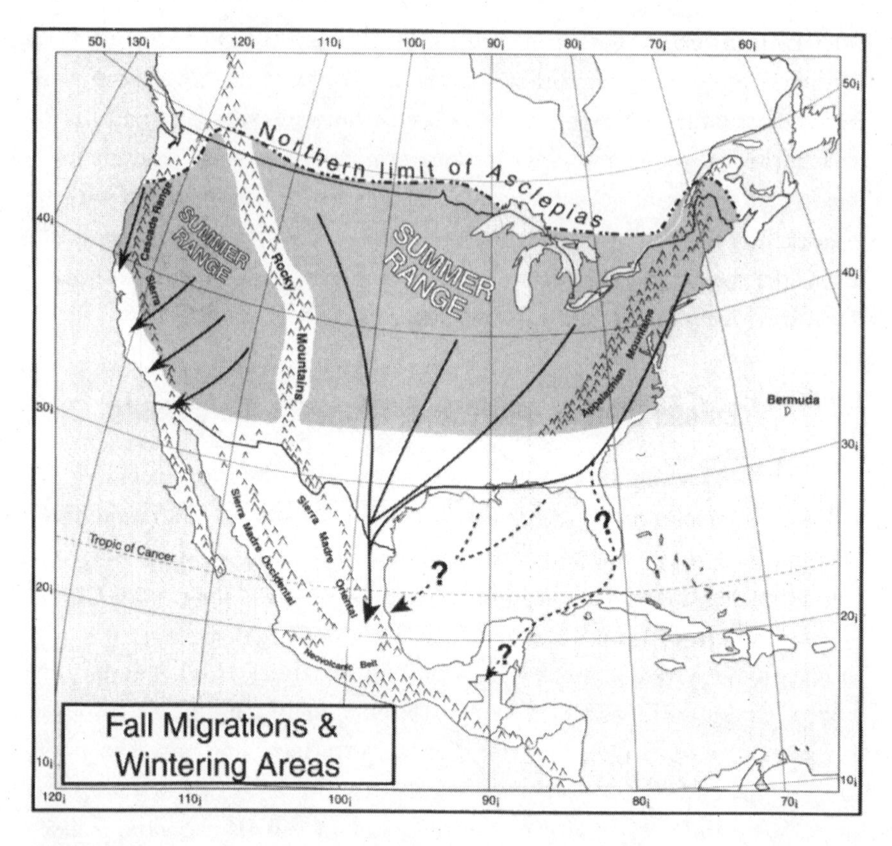

Figure 8.1. The traditional model of autumnal migration of monarchs in North America (from Brower 1995a).

model and whether it fully expresses the reality of western monarch movement (W. Sakai, G. Austin, and J. Dayton, personal communication). Reexamination of the basis for the East-West model has shown that it was predicated on very few actual data (Pyle 1999), and as this chapter will show, the degree of interchange appears more extensive than the body of prior evidence indicates. For decades, the tag-release-recapture studies had pointed the way toward Mexico for the eastern monarchs, and this proved to be an accurate prediction. But in the West, little tagging had been conducted outside California, and very few tag recoveries were made. Of those that were tagged and released in the West, most consisted of monarchs that had been captured at various California and Ontario locations and then shipped by mail to sites in Idaho, British Columbia, and elsewhere. Once released, some of these artificially transferred monarchs were recovered in California (Urquhart and Urquhart 1977). For example, in 1972 Donald Davis (pers. comm.) sent 900 pre-tagged monarchs from Ontario, in lots of 50, to Maryanne West of Gibsons, British Columbia. Gibsons is north of Vancouver, more than 200 km west of the nearest native milkweed stands. According to Davis, P. Cherubini later found two of these butterflies in northern California. Such are the data upon which large-scale generalizations about western monarch movements have been based.

Cherubini transferred and released a large number of monarchs east and west of the Continental Divide in the fall of 1999 (Monarch Watch, dplex-1@raven.cc.ukans.edu, February 7, 2000). Some individuals reputedly were recovered along the California coast and others at the known overwintering sites in Mexico. Although the report has not been available for critical review and the methods have not been published, the data, if true, support our hypothesis that monarchs situated west of the Continental Divide are capable of navigating to Mexico and those east of the Continental Divide to the California coast. However, we consider this exercise deplorable for the reasons given below, of questionable legality, and useless for making broader generalizations.

Such transfer schemes are scientifically and empirically flawed in fatal ways. First, it is fallacious to assume that manipulated behavior in one location can be generalized to demonstrate natural behavior elsewhere. Thus, although a monarch from site A, released at site B, and recovered at site C may demonstrate that individuals are capable of getting from release site B to the recovery site C, the exercise does not show that monarchs originating naturally at site B *ever actually fly* to site C. Second, given the various physiological changes that take place under differing light and temperature regimes associated with latitudinal or longitudinal shifts, as well as the influences of differing terrains, biological interpretations drawn from these artificial transfers seem moot. Transfers from the eastern to the western states would probably exaggerate these effects, and, if the originating and recipient populations differ genetically (see Eanes and Koehn 1978; Brower et al. 1995; Altizer 1999), the butterflies may experience even more severe potential effects.

Despite the questionable utility of transfers in illuminating the migratory patterns of unmanipulated monarchs, conclusions based on these transfer "experiments" have been accepted by serious monarch students, including W. Sakai and D. Marriott (pers. comm.) in the mistaken belief that they represent natural migratory flights. A thorough search of available data (Pyle 1999) revealed only two groups of monarchs that were tagged and released in their natal habitat and then recovered in California. Both groups were reared in the classroom from locally collected larvae and tagged and released by two Boise, Idaho, schoolteachers, Faye Sutherland and Mary Henshall (pers. comm.). Most of the recoveries, about twenty in all, occurred at several locations in central and southern California. But what has generally been ignored is the fact that two monarchs tagged by these women in the same location were also recovered far to the east and southeast, in Orem and St. George, Utah. Although the samples are small, they are notable when one considers that many observers watch for returning monarchs in California, but almost none do so in the Great Basin; yet the portion of the total Boise sample found far to the southeast is some 10 percent of the number found to date in California.

More recently, a wild adult monarch captured by D. Branch and tagged by R. Pyle at Roosevelt, Washington, in the Columbia Gorge, was recovered in Aptos, near Santa Cruz, California. Released on September 26, 1997, and recaptured on October 26 by J. Lovenfosse, this was the first wild-captured adult monarch from any other western state to be shown to migrate to California (Pyle 1997). But many other monarchs from the Columbia Basin were found to depart the region southeasterly (see below, and table 8.1). In view of these observations, and given the paucity of overall numbers in the West, the heterogeneous ways in which butterflies were handled prior to release, and the broad spread of the few recoveries, we conclude that the generally assumed southwesterly direction of fall migrants west of the Rockies is poorly supported by the available data. We therefore present two lines of recent evidence suggesting monarch exchange between Mexico and the western United States, and their implications.

Recent Observations of Southeasterly Movements in the Western United States

In an effort to determine whether the traditional maps or a different model better describes the facts in the field, monarch emigration from the

TABLE 8.1. Observations of Autumnal Monarch Movements in the West

Date	Location	Number	Flight Direction
2 Sept	Cawston, Similkameen R, BC	1	S
9 Sept	Lenore Lake, Lower Grand Coulee WA	1	S
13 Sept	Columbia R ab Vernita Bridge WA	1	S
15 Sept	Columbia R Hanford Reach WA	4	2E, SE, S
15 Sept 97	Columbia R Hanford Reach WA	4	2ESE, E, S
16 Sept	Columbia R at Kennewick WA	1	SE
17 Sept	Columbia R at Pasco WA	2	SW, SSE
17 Sept	Columbia X Snake R WA	1	SSW
18 Sept	Columbia X Walla Walla R WA	1	E
18 Sept	Columbia River at McNary Dam OR	1	W
27 Sept 97	Columbia R nr Irrigon OR	9	3ESE, 2E, NE, SE, NNE
28 Sept 97	Birch Crk W of Pilot Rock OR	3	SW, S, W
20 Sept	Columbia R at Crow Butte WA	1	W
20 Sept	Columbia R at Roosevelt WA	1	W
26 Sept 97	Columbia R at Roosevelt WA	2	SW, ESE
30 Sept 97	Columbia R at Maryhill WA	1	SE
12 Oct 97	Columbia R at Maryhill WA	3	E, S, ESE
28 Sept 97	Columbia R X John Day R OR	4	3ESE, SE
5 Sept 97	Murderers Crk, John Day R OR	2	E, SE
23 Sept	Snake R at Brownlee Res ID	7	S
28 Sept	Snake R at Big Bar, Hell's Cyn ID	4	S
1 Oct	Snake R at 1000 Springs ID	1	ESE
4 Oct	Bear R at Corinne UT	1	S
5 Oct	S of Snake R at Grasmere, ID	1	SE
5 Oct	N of Adobe Pass ID/NV	1	SE
6 Oct	Bonneville Salt Flat, I-70, UT	1	ESE
7 Oct	Clear Lake Refuge, UT	1	SSE
18–21 Sept	Desolation Canyon, Green R UT M. Monroe (personal communication)	30+	S
16 Sept	Confluence Little Colorado R and Colorado R, Grand Canyon AZ G. Nabhan (personal communication)	5–600	S
12 Oct	Salt River Cyn AZ	1	W
12 Oct	N of Globe AZ	3	2NNE, WNW*
12 Oct	San Carlos R, Apache Res. AZ	1	S
12 Oct	N of Wilcox AZ	2	W
13 Oct	Pinery Cyn, Chiricahua Mtns AZ	1	SSE
14 Oct	E of Douglas AZ	3	2S, SE

*Caught up in mass northward movement of snout butterflies.

Source: All records are for 1996 from Pyle (1999) unless otherwise indicated. Many more individuals were sighted; these data include only those individuals demonstrating a discrete vanishing bearing, as opposed to milling, surface foraging, or roosting flight.

western states and provinces was observed by Pyle in the autumn of 1996. The results were reported along with other anecdotal information gathered from a variety of sources (Pyle 1999). Beginning in the northwesternmost breeding grounds in the Okanagan Valley in British Columbia and following successive vanishing bearings taken on individual migrant monarchs, Pyle observed that the butterflies appeared to fly southeasterly out of the inland Northwest and the Great Basin more than they flew southerly or southwesterly, and that migratory movement was evident along the Okanogan (WA), Columbia (WA/OR), and Snake (ID) rivers. During the same period, M. Monroe and G. Nabhan reported substantial movements along the Green (UT) and Colorado (AZ) rivers, respectively (pers. comm.). Ultimately, on October 14, 1996, Pyle found three monarchs strongly migrating south and south-southeast, five km north of the Mexican border, about 100 km west of the Continental Divide, between San Bernardino National Wildlife Refuge and Guadalupe Canyon in southeastern Arizona. These data have been summarized in table 8.1.

Confirming records of autumn monarchs moving southward in Arizona were reported by several independent field observers (all personal communication, discussed in Pyle 1999). R. Hanson, C. Melton, H. Brodkin, and R. Pyle recorded several instances of autumnal monarch movement toward Mexico along the San Pedro River riparian corridor, and others in the Buenos Aires National Wildlife Refuge southwest of Tucson. A businesswoman interviewed by Pyle convincingly described an overnight aggregation of hundreds of monarchs on mesquite in September 1996 at Sasabe, 2 km from the Mexico border, that were gone the next day. And in early October 1998, R. and E. Gill observed twelve monarchs in and near Organ Pipe Cactus National Monument, four of which were flying with a definite southern orientation on October 5. The Gills watched one of these crossing the border heading south, 5 km west of the junction of the main park road along the Sonoita Basin Road near Quitobaquito Springs. And a historical narrative of a pioneer child in the Arivaca Valley very near the Mexican border in southern Arizona (Wilbur-Cruce 1991) tells of encountering large numbers of migrating monarchs in an overnight bivouac. These data are summarized in table 8.2.

TABLE 8.2. Arizona Monarch Movements near the Mexican Border

Date	Location	Number	Conditions	Source
Early 1900s	Arivaca Valley	"large numbers"	covering trees	Wilbur-Cruce (1991)
1996–99	Brown Cyn, Buenos Aires NWR, 30 km N of border	many	nectaring, breeding, flying through	R. Hanson (pers. comm., 1999)
Sept 96	La Osa Bar, Sasabe 2 km N of border	1–300	bivouac in mesquite, depart to S	Pyle (1999)
13 Oct	Pinery Cyn, Chiricahua Mtns, 70 km N of border	1	flying to SSE	Pyle (1999)
14 Oct 96	25 km E of Douglas 5 km N of border	3	flying to S, SE	Pyle (1999)
June 97	St. David Cienega San Pedro R., 50 km N of border	"many"	nectaring on button willow	R. Hanson (pers. comm., 1999)
3 Nov 97	San Pedro R. NCA, (riparian reserve) 20 km N of border	1	flying upriver to S	R. Hanson (pers. comm., 1999)
2–4 Oct 98	Organ Pipe NM and Ajo, 0.5 to 50 km N of border	8	random flight	R. and E. Gill (pers. comm., 1999)
5 Oct 98	Organ Pipe NM Bonita Well, Aguajita Sprg, Quitobaquito Spring, 0.5 to 10 km N of border	3	flying to S	R. and E. Gill (pers. comm., 1999)
5 Oct 98	Organ Pipe NM Sonoita Basin Rd. at border	1	crossing border >S	R. and E. Gill (pers. comm., 1999)
Sept 1999	San Pedro E of Sierra Vista, 30 km N of border	several	moving south	Pyle (1999)
20 Aug–23 Sept 00	Hereford 14 km N of border	5	nectaring on Cosmos, Tithonia	C. Melton (pers. comm., 2000)
5 Sept 00	Carr Canyon 16 km N of border	1	flying	H. Brodkin (pers. comm., 1999)
late Oct 00	Fort Huachuca 30 km N of border	several	nectaring	C. Melton (pers. comm., 1999)

Do Some Western Migrants Overwinter in Mexico?

The destination of these western monarchs flying into Mexico during the fall is unknown. A modest amount of monarch overwintering takes place along the west coast of Baja California, within 90 km of the U.S. border. Marriott (1999) reported that the dominant movement of autumn monarchs in Baja is northerly, but he also reported some movement from Baja to Arizona (Ensenada to Gila Bend), northwesterly across the Sea of Cortés from mainland Mexico, and possibly trans-Baja from southeast to northwest. He estimated that 85 percent of Baja wintering monarchs depart to the north, and speculated (pers. comm.) that the Arizona border-crossers in the fall probably end up in Baja. However, this would require a sharp westerly right turn, which seems unlikely to us.

Another possibility is that the migrants continue southeasterly across the Sonoran Desert into the Sierra Madre Occidental and overwinter in the high montane forests in the state of Durango (see Carta de uso del suelo y vegetacion 1981a, 1981b). Alternatively, they may continue flying southeasterly across the Tropic of Cancer to the known overwintering sites of eastern monarchs in the *Abies religiosa* forests in the Transverse Neovolcanic mountains west of Mexico City (Carta 1981c; Calvert and Brower 1986). Following these fall migrants beyond Arizona into Mexico would be a worthwhile undertaking along the lines of Pyle's 1996 extended field project. This further pursuit could determine once and for all whether additional overwintering areas exist in Mexico, outside the Neovolcanic zone. Regardless of their ultimate destination, it is now clear that at least some monarchs from west of the Continental Divide migrate into mainland Mexico.

Other Approaches Bearing on the East–West Question

Because the Rocky Mountains do not constitute the barrier they were previously assumed to be, more interchange between the eastern and western monarchs may take place than heretofore realized. Four separate research approaches could bear upon the degree of interchange and therefore the genetic homo- or heterogeneity of the eastern and western populations. First, Malcolm et al.'s (1993) cardenolide fingerprint analyses of several hundred monarchs overwintering west of Mexico City indicated that most had the fingerprint of the eastern milkweed, *Asclepias syriaca*, with few other fingerprints. If some monarchs had migrated from California, they could have had distinct fingerprints of other milkweed host species, such as *Asclepias eriocarpa, A. cordifiolia, A. vestita*, or *A. erosa* (Roeske et al. 1976). None did. However, a potential confounding factor is that the western *A. speciosa* is a major food plant throughout the Northwest (Pyle 1999) and the Rocky Mountain states (Woodson 1954), and its fingerprint is difficult to distinguish from that of the eastern *A. syriaca* (Malcolm et al. 1989). Thus the fingerprinting evidence for western monarchs moving during the fall migration period into the eastern population was negative, but weak.

A second approach (Brower and Boyce 1991) analyzed mitochondrial DNA from monarchs collected in Trinidad, W.I., and from overwintering sites in Mexico and California. It found virtually no differences among the three populations. However, the sample sizes were small and few markers were sampled, so that more extensive research is needed using this and other molecular approaches.

Third, Wassenaar and Hobson (1999) compared hydrogen isotopes in Mexico-wintering monarchs with defined profiles from U.S. groundwater sources and milkweeds, which led to the conclusion that most monarchs at the overwintering sites west of Mexico City had bred in the U.S. corn belt east of the Rocky Mountains. However, because no western plant samples were included in their deuterium baseline, the East versus West question remains moot with respect to this experimental approach.

A fourth approach (Altizer 1998; Altizer et al. 2000) was to analyze the susceptibility of eastern and western monarchs to infection by the parasitic schizogregarine protozoan, *Ophryocystis elektroscirrha*. Altizer determined that western monarchs were more susceptible to infection, particularly by the protozoans from their native populations. Eastern monarchs inoculated with eastern parasites had the lowest parasite loads as emerging adults, and western monarchs inoculated with western parasites had the highest loads (of all four possible combinations). Of particular significance was the finding that the western parasites were more infectious and caused higher pre-adult mortality than the eastern parasites. Altizer's work also suggested the possibility of significant morphological differences between monarch adults from the two populations, including wing length (see also Arango-Velez 1996).

A New Hypothesis on Occasional Major East to West Influxes

Independently, L. Brower and S. Gauthreaux (Brower 1999a) speculated in 1996 that there may be occasional major influxes of eastern monarchs into the western range, and they proposed a new hypothesis that the long-term survival of the western population may ultimately depend on these large influxes. Four lines of evidence were adduced to support this hypothesis:

1. Beginning in 1992, the numbers of monarch butterflies at numerous overwintering sites along the coast of California began to decline precipitously and during the 1995–96 season were estimated to be at an all-time low (Cherubini 1995; Frey 1995; Sakai 1995). Annual summer censuses corroborated the declining western population (Swengel 1994). A possible reason for this was an increasing incidence of the above-mentioned protozoan disease that debilitates adult monarchs: collections from several California overwintering sites determined that up to 66 percent of the adults were infected (Leong et al. 1997; Altizer et al. 2000).

2. Monitoring the 1996 spring remigration of the eastern population indicated that the numbers of monarchs returning from Mexico were far below normal. Thus, monarchs captured during April 1996 in northern Florida were down by nearly 90 percent from the numbers estimated the previous spring (Knight 1998; Knight et al. 1999). Observations during the same period in central Texas also indicated a severe reduction from the usual numbers (Calvert, in Taylor 1996).

3. The numbers of first-generation monarchs born in the south that migrate to the Great Lakes region were also severely depressed: compared to 1994 and 1995, the numbers arriving in northern Wisconsin in June 1996 were down by more than 90 percent (Brower and Borkin, in Davis 1996).

4. Paradoxically, by mid-summer 1996 monarchs were reported to be abundant at several California locations (Cherubini 1996; Marriott 1996). By November the butterflies clearly had made a major recovery, as indicated by large numbers returning to overwintering colonies at Pacific Grove and southward (Arnott 1996; Vaccaro 1996).

How could the California monarchs spring back so suddenly? Brower and Gauthreaux hypothesized that the poor showing of spring monarchs in the eastern population and their sudden recovery in California during the summer of 1996 may both have been a consequence of a major westward displacement of monarchs during their 1996 spring remigration from the Neovolcanic overwintering area in Mexico. This purported shift was correlated with a major westward shift in the normal spring wind patterns along the Gulf Coast. Their hypothesis was supported by radar and field observations that indicated a major western displacement of the spring warbler migration during March and April 1996—that is, at virtually the same time that monarch butterflies remigrate northward from the known Mexican overwintering areas. Thus, rather than migrating in a broad band from the Mississippi Valley to the east coast, the eastern edge of the warbler migration seemed to have been displaced more than 2,400 km to the west (Gauthreaux, pers. comm.). If the monarchs returning in the spring from Mexico were similarly blown westward off their normal course, then the resulting displacement may have dispersed many eastern monarchs across Mexico and Texas into New Mexico and Arizona, which may have then resulted in the California population recovery. In addition, Brower and Gauthreaux hypothesized that such a severe western displacement might also have accounted for the monarchs' extreme paucity during the spring and early summer in Texas, Florida, and Wisconsin.

Implications of Western U.S.–Mexico Monarch Exchange

These combined observations suggest that the old model fails to describe the whole of the monarch movement in the West. Figure 8.2 illustrates an alternative model, expressed in terms of hypothetical spring remigration from Mexico and California. Although details remain uncertain, the fact seems inescapable that the western population contributes autumn emigrants to both Mexico and California, and in turn receives spring immigrants from both regions as well. How the western states' monarchs divide and interact must be the subject of further research.

Of more immediate importance, these arguments also suggest that the long-term survival of the West Coast population of monarchs may depend on occasional replenishment from the Mexican winter population. If

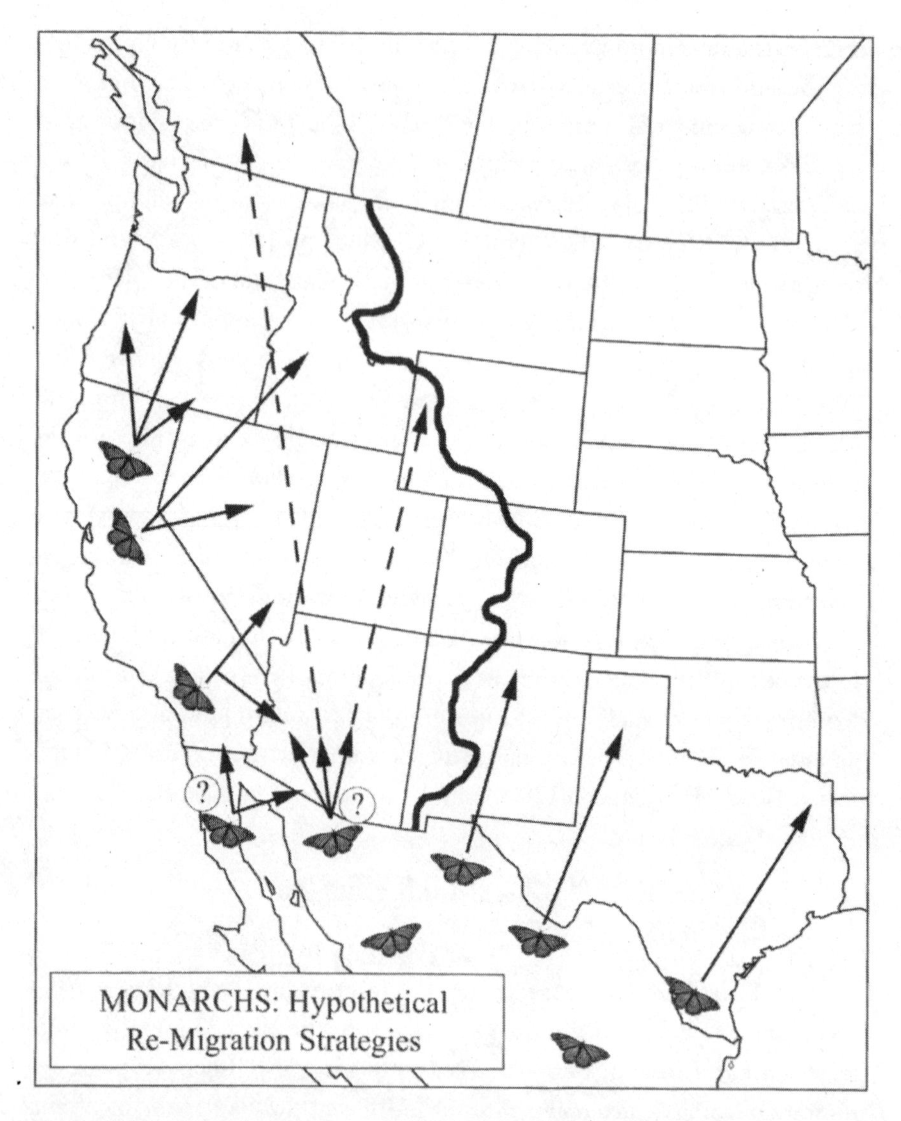

Figure 8.2. Potential remigration strategies in the spring, indicating the Arizona-Mexico and Mexico-California exchanges discussed in this chapter.

the Brower-Gauthreaux hypothesis proves correct, then both the eastern and western North American monarch populations and their migrations are ultimately dependent on the Mexican winterers. Under this scenario, continued deforestation in Mexico could result in total collapse of the entire North American migratory phenomenon. As Pyle (2001) stated in a response to the erroneous *Sunset* map, referring to the findings reported here, "The significance of this discovery . . . is that the Californian and Mexican monarchs cannot be saved piecemeal, but are intrinsically linked."

Unfortunately, the threat to the Mexican phenomenon has increased dramatically in recent years (Snook 1993; Anderson and Brower 1996; Aridjis and Brower 1996; Merino-Perez and Gerez-Fernandez 1996; Brower and Missrie 1999). Brower et al. (2002) carried out a Geographic Information System study indicating that the oyamel-pine forest ecosystem in which the entire eastern population of the monarch butterfly overwinters is being degraded faster than we thought. Between 1971 and 1999, 44 percent of the high-quality forest in the 27,485 ha study area was degraded. Thus, what in 1971 was nearly continuous high-quality forest is now a series of smaller fragments with many patches of degraded forest between them. The annual rate of degradation was 1.70 percent from 1971 to 1984 and this increased to 2.41 percent over the next fifteen years. At this rate, there will be fewer than 10,000 ha of high-quality forest remaining in twenty years, and fewer than 4,500 ha in fifty years. The corresponding rates of degradation in the supposedly protected Sierra Chincua, Sierra Campanario, and Chivati-Huacal reserves more than tripled, from 1.03 to 3.17 percent per annum.

All indications are that the rate of forest degradation overall is increasing. Brower et al.'s study provided irrefutable evidence that passage of the 1986 presidential decree failed to result in effective protection and that the new presidential decree issued in November 2000 (Zedillo 2000) must (a) ensure effective enforcement against illegal logging, (b) follow management guidelines that ensure the integrity of the oyamel-pine forest ecosystem, and (c) restore areas within the new reserve that have been degraded. In light of our findings linking western states' and Mexican monarchs, preservation of these Mexico forests takes on an even greater urgency than has already been appreciated.

Caveat on Interpopulational Transfers

Although our new evidence suggests that there may well be interchange between the eastern and western populations, we still have no idea how often or to what extent this may occur. Furthermore, because major questions remain concerning the genetic and biogeographic integrity of the migratory populations, we urge other workers to reject ex situ monarch releases in general, so as not to generate additional spurious distributional data. Monarchs released far from their natural points of origin, whether for events such as weddings and funerals or for ill-defined and misguided transfer "experiments," hold the potential for permanently confusing our broad understanding of monarch movements in the West (Pyle 1998, 1999) and subjecting the two populations to genetic muddling and to possible infection from reciprocally introduced diseases (Brower et al. 1995).

More generally, given that the monarch's migration and overwintering behavior is an endangered biological phenomenon (Brower and Pyle 1980; Pyle 1983c; Wells et al. 1983; Brower and Malcolm 1991; Brower 1997), we strongly urge researchers and commercial vendors to desist from activities that will further confuse the still poorly understood relationships between the eastern and western populations in North America. To do otherwise is to jeopardize an intelligent approach to conservation of the endangered monarch phenomenon.

Migratory Corridors and the Importance of Preserving Their Floral Resources

The various lines of evidence presented here demonstrate that migratory corridors probably do occur between Arizona and Sonora and between Alta and Baja California. At least the Guadalupe, San Pedro, Buenos Aires, and Organ Pipe corridors are used by monarchs (see table 8.2 and figure 8.2). Further, monarchs are just one of many species of butterflies and moths that utilize these corridors. Although the other lepidopterans are largely if not completely emigratory rather than round-trip migrants like monarchs, they perform significant pollination services and must require copious nectar during their peak seasons and years. In fact, nectar availability in the spring has been put forth by various authors as a determining factor for major irruptions of these species. Among the more irruptive species in a given spring are several species of coliadines such as the Mexican sulfur *(Eurema mexicana)*, the dainty sulfur *(Nathalis iole)*, and the cloudless giant sulfur *(Phoebis sennae)*; lycaenids including the marine blue *(Leptotes marina)* and Reakirt's blue *(Hemiargus isola)*; several nymphalids, notably the painted lady *(Vanessa cardui)*, the red admiral *(V. atalanta)*, and the variegated fritillary *(Euptoeita claudia)* as well as the white-lined sphinx moth (*Hyles lineata*; Pyle 1981; Scott 1986).

The importance of nectar plants along the migratory corridors has been addressed by García and Equihau-Zamora (1997) and can be intuited by observing the large numbers of butterflies and moths nectaring on such autumn-blooming shrubs as rabbitbrush *(Chrysothamnus nauseosus)* and desert broom (*Baccharis* spp.). Other genera frequently employed by monarchs during their western migration include *Aster, Helianthus, Solidago, Verbena, Trifolium*, and *Asclepias* (Pyle 1999). William H. Calvert (pers. comm.) reports having seen monarchs nectaring extensively during the fall migrations on frostweed (*Verbesina virginica*, Compositae), which occurs as a major component of the understory floodplain flora in central and western Texas riverine ecosystems. Other nectar sources he has found important for monarchs in central Texas include gayfeather *(Liatris mucronata)* and various species of *Eupatorium*, including white boneset *(E. serotinum)*, blue mist-flower *(E. coelestinum)*, and thoroughwort *(E. havanense)*. In northern Mexico, Calvert considers the chief nectar plants during migration to be Mexican sunflower (*Tithonia* spp.), cowpen daisy *(Verbesina encelioides)*, and golden-eye *(Viguera dentata)*.

The extreme lipid-loading of monarchs during the southward migration through Texas and northern Mexico (Brower 1985; Alonso-Mejia et al. 1997) is indisputable evidence for the importance of the corridor flora for the eastern population. A significant proportion of monarchs that arrive in Mexico lack sufficient lipids to survive the winter. Brower, O'Neil, and Missrie (unpublished data) have found that virtually all monarchs attempting to nectar on species of *Senecio* and *Lupinus* in the Sierra Chincua area are in fact starving with no chance of obtaining significant sugar on the desiccated and grossly depleted flowers near the overwintering colonies (see also Brower 1995b as contrasted to Hoth 1995). However, after the monarchs leave the overwintering sites, nectar sources are probably again crucial. In fact, extensive nectar-drinking during the spring remigration in Mexico has been reported by Heitzman (1962).

It is clear that herbicides, agriculture, development, and any other factors that reduce available nectar resources along the migratory pathways and corridors will have deleterious effects on monarchs and their ability to reach their winter clusters and summer breeding grounds (Brower 1999b).

Because monarchs and other butterfly migrants use river courses and other linear pathways extensively (Pyle 1999), a conservation strategy to aid migrants would include promotion of unsprayed nectar-rich riparian zones and roadsides. Highway verge spraying, close and ill-timed mowing, and denuding of floodplain vegetation all work against monarchs and other butterfly migrants.

Encouragement of nectiferous river valleys, roadsides, neighborhoods, parks, gardens, and public lands would furnish enormous benefits to Lepidoptera of passage. The importance of resource oases was emphasized by Russell et al. (1994) based on their work with rufous hummingbirds. They emphasized that an understanding of the variability in stopover habitats is important to addressing questions of conservation and loss of such places. They make the vital point that fragmentation and degradation of stopover habitats can render them unusable by migrants and can reduce the birds' chances of completing migration. This is certainly true for butterflies as well.

In both authors' experience, monarchs appear to descend late in the day into sites hospitable not only for overnight bivouac but also for nectaring before the night roost and prior to morning departure. Loss of nectar resources could have the practical effect of rendering such resource oases unsuitable. Monarchs also frequently nectar between high flights, after reaching the bottom of a glide and before rising again on a thermal. As Brown and Chippendale reported (1974: 1128), "Although the food consumption of monarch butterflies during the fall migration is regulated by temperature and available flowering plants, they consume nectar whenever conditions permit." This supports the notion that patchy nectar sources may be as important as concentrated stands of preferred plants.

Although migratory monarchs and other Lepidoptera follow corridors facultatively, they also conduct broad-front cross-country movement when weather, geographic, or habitat constraints dictate. Therefore, the use of biocides in the open countryside, affecting both nectar plants and the adults directly, interferes with migrations. Of course any factors deleterious to milkweed growth (Cuddeford 1999) diminish the ability of returning spring monarchs to reestablish their summer breeding generations in the north. In this respect, herbicides and genetically engineered herbicide-resistant crop species are particularly pernicious, not only to the monarch but to the entire biota as well (Brower 1999b).

Future research should examine the magnitude of lepidopteran use of nectar corridors and the array and frequency of plant species utilized. Optimal management and conservation strategies for nectar and host-plant vitality in the critical areas may entail support of traditional versus intensive forms of agriculture (New et al. 1995). Traditional agriculture can also promote greater nectar availability than no agriculture, as described by Nabhan (1987) for Quitobaquito Spring in Arizona—the same area where a western monarch was first observed actually crossing the Mexican border. We wish to emphasize that the movement dynamics and nectaring needs of monarch butterflies and all other migrants are undoubtedly very different during the fall and spring migrations.

Conclusion

Although most of the evidence accrued to date suggests that the eastern and western populations are largely allopatric or parapatric, the new observations assembled in this chapter suggest that the Rocky Mountains may not constitute the virtually complete barrier they were previously assumed to be. A new model of western monarch movements will therefore have to take into account the likelihood of some exchange of adults between western breeding habitats and overwintering sites in Mexico, and the reverse. We know that some monarchs arrive at the California aggregations from the Columbia and Snake River basins, but it is likely that the bulk of the Pacific coastal colonies breed within California, and that autumn migrants from other western states travel both southeasterly toward Mexico and southwesterly to the California coast. A new model must also address the possibility that monarchs that overwinter in the Transverse Neovolcanic Belt of Mexico and possibly also in the Sierra Madre Occidental of western Mexico contribute to the West Coast population, which fluctuates dramatically and may, over the long term, be dependent on the well-being of these Mexican winter roosts.

These revised views of monarch migration in western North America show that their movements and conservation may be much more complex than previously supposed. They also suggest that monarchs (and other transborder butterfly and moth immigrants, of which there are many species, though none are known to have the monarchs' dramatic two-way migration) may play a considerable role in pollination systems in the Southwest, and that, to reach their destinations, these butterflies depend on nectar resources on both sides of the border.

NOTE IN PRESS

On August 2, 2003, the Ramsey Canyon Butterfly Count (conducted in and near southern Arizona's Huachuca Mountains) found 65 monarchs along the San Pedro River between Hereford and Palominos (H. Brodkin, pers. comm.). The site supported a robust stand of *Asclepias subverticillata*. This adds further strong evidence for a transborder movement between Arizona and Sonora.

ACKNOWLEDGMENTS

The research on which Brower's contribution to this chapter is based has been supported since 1977 by the U.S. National Science Foundation, Amherst College, the University of Florida, Sweet Briar College, the World Wildlife Fund, the Wildlife Conservation Society, and two private donors. He thanks his many colleagues and students for their dedication to his monarch butterfly research program. Pyle's investigations were supported by the Houghton Mifflin Company of Boston. He is grateful to the many people who helped with the field work, logistics, and literature search and who contributed supportive observations and refining argumentation, most of whom are named here or in *Chasing Monarchs: Migrating with the Butterflies of Passage*. Kind thanks are due to Lauri Walz for figure 8.1 and David White for figure 8.2. We both thank William H. Calvert in particular for his many contributions toward understanding monarchs and for his helpful nectaring observations.

LITERATURE CITED

Ackery, P. R., and R. I. Vane-Wright. 1984. Milkweed Butterflies: Their Cladistics and Biology. Cornell University Press, Ithaca, N.Y.

———. 1985. Patterns of plant utilization by danaine butterflies. Proceedings of the Third Congress of European Lepidoptera, Cambridge, England (1982), pp. 3–6.

Alonso-Mejia, A., E. Rendon-Salinas, E. Montesinos-Patino, and L. P. Brower. 1997. Use of lipid reserves by monarch butterflies overwintering in Mexico: Implications for conservation. Ecological Applications 7:934–47.

Altizer, S. M. 1998. Ecological and evolutionary interactions between monarch butterflies and the protozoan parasite, *Ophryocystis elektroscirrha*. Ph.D. diss., University of Minnesota.

———. 1999. Letter re interchange of monarchs across the continental divide. Monarch Watch (dplex-1@raven.cc.ukans.edu), attachment to Report 1217, July 9.

Altizer, S. M., K. S. Oberhauser, and L. P. Brower. 2000. Association between host migration and the prevalence of a protozoan parasite in natural populations of adult monarch butterflies. Ecological Entomology 25:125–39.

Anderson, J. B., and L. P. Brower. 1996. Freeze-protection of overwintering monarch butterflies in Mexico: Critical role of the forest as a blanket and an umbrella. Ecological Entomology 21:107–16.

Arango-Velez, N. 1996. Stabilizing selection in migratory butterflies: A comparative study of queen and monarch butterflies. Master's thesis, University of Florida, Gainesville.

Aridjis, H., and L. P. Brower. 1996. Twilight of the monarchs. The New York Times, January 26, p. A-15.

Arnott, C. L. 1996. Current census reports. Internet, Monarch Watch (dplex-1@raven.cc.ukans.edu), December 11, 9:45 P.M.

Associated Press. 2000. Mexico to pay to conserve forests. November 10.

Brower, A. V. Z., and T. M. Boyce. 1991. Mitochondrial DNA variation in monarch butterflies. Evolution 45:1281–86.

Brower, L. P. 1985. New perspectives on the migration biology of the monarch butterfly, *Danaus plexippus* L. Pp. 748–85 in M. A. Rankin, ed., Migration: Mechanisms and Adaptive Significance. University of Texas, Austin.

———. 1995a. Understanding and misunderstanding the migration of the monarch butterfly (Nymphalidae) in North America: 1857–1995. Journal of the Lepidopterists' Society 49:304–85.

———. 1995b. Revision a los mitos de Jurgen Hoth (Jurgen Hoth's myths revisited). Ciencias No. 39 (July–September):50–51.

———. 1996. Monarch butterfly orientation: Missing pieces of a magnificent puzzle. Journal of Experimental Biology 199:93–103.

———. 1997. A new paradigm in biodiversity conservation: Endangered biological phenomena. Pp. 115–18 in G. K. Meffe and C. R. Carroll, eds., Principles of Conservation Biology. Sinauer, Sunderland, Mass.

———. 1999a. Oyamel forest ecosystem conservation in Mexico is necessary to prevent the extinction of the migratory phenomenon of the monarch butterfly in North America. Pp. 41–50 in U. C. Secretariat and P. Canevari, eds., Proceedings of the CMS Symposium on Animal Migration, Gland, Switzerland (April 13, 1997). United Nations Environment Programme, Convention on the Conservation of Migratory Species of Wild Animals (UNEP/CMS), Bonn, The Hague.

———. 1999b. Will biotechnology doom the monarch? Defenders 79:39–41.

Brower, L. P., G. Castilleja, A. Peralta, J. Lopez-Garcia, L. Bojorquez-Tapia, S. Diaz, D. Melgarejo, and M. Missrie. 2002. Quantitative changes in forest quality in a principal overwintering area of the monarch butterfly in Mexico: 1971 to 1999. Conservation Biology 16:346–59.

Brower, L. P., L. S. Fink, A. V. Z. Brower, K. Leong, K. Oberhauser, S. Altizer, O. Taylor, D. Vickerman, W. H. Calvert, T. Van Hook, A. Alonso-M., S. B. Malcolm, D. F. Owen, and M. P. Zalucki. 1995. On the dangers of interpopulational transfers of monarch butterflies. Bioscience 45:540–44.

Brower, L. P., and S. B. Malcolm. 1989. Endangered phenomena. Wings 14:3–10.

———. 1991. Animal migrations: Endangered phenomena. American Zoologist 31:265–76.

Brower, L., and M. Missrie. 1999. State of the monarchy in Mexico: An early update on the 1998–99 season. Monarch News 9(1):3–4.

Brower, L. P., and R. M. Pyle. 1980. Remarks on endangered wildlife spectacles. Proceedings of the 54th Meeting of the International Union of the Conservation of Nature and Natural Resources Survival Commission, Gainesville, Fla., pp. 1–26.

Brown, J. J., and G. M. Chippendale. 1974. Migration of the monarch butterfly, *Danaus plexippus*: Energy sources. Journal of Insect Physiology 20:1117–30.

Calvert, W. H., and L. P. Brower. 1986. The location of monarch butterfly (*Danaus plexippus* L.) overwintering colonies in Mexico in relation to topography and climate. Journal of the Lepidopterists' Society 40:164–87.

Carta de uso del suelo y vegetacion 1:1,000,000. 1981a. Estados Unidos Mexicanos, Guadalajara. Direccion General de Geografia del Territorio Nacional SPP, México D.F.

———. 1981b. Estados Unidos Mexicanos, Chihuahua. Direccion General de Geografia del Territorio Nacional SPP, México D.F.

———. 1981c. Estados Unidos Mexicanos, México. Direccion General de Geografia del Territorio Nacional SPP, México D.F.

Cherubini, P. 1995. The California population crash. The Monarch Newsletter 5(7):7.

———. 1996. A great year for western monarchs. The Monarch Newsletter 6(10):1, 3.

Cuddeford, V. 1999. Beneficial Bugs at Risk from Pesticides. World Wildlife Fund Canada, Toronto, Ontario.

Davis, D. 1996. Low numbers east of the divide. The Monarch Newsletter 7(6):3.

Eanes, W. F., and R. K. Koehn. 1978. An analysis of genetic structure in the monarch butterfly, *Danaus plexippus* L. Evolution 32:784–97.

Frey, D. 1995. California overwintering sites. The Monarch Newsletter 5(10):6–7.

García, E. R., and M. E. Equihau-Zamora. 1997. New records of plant species used by adult monarch butterflies *Danaus plexippus* L. (Lepidoptera: Nymphalidae: Danainae) during migration in Mexico. The Canadian Entomologist 129:375–76.

Heitzman, R. 1962. Butterfly migrations in March in northern Mexico. Journal of the Lepidopterists' Society 16:249–50.

Hoth, J. 1995. Mariposa monarca, mitos y otras realidades aladas. Ciencias 37:19–28.

Jenson, P. 2000. Mapping the West's monarch migration. Sunset (November):46.

Knight, A. L. 1998. A population study of monarch butterflies in north-central and south Florida. Master's thesis, University of Florida, Gainesville.

Knight, A. L., L. P. Brower, and E. H. Williams. 1999. Spring remigration of the monarch butterfly, *Danaus plexippus* (Lepidoptera: Nymphalidae) in north central Florida: Estimating population parameters using mark-recapture. Biological Journal of the Linnean Society 68:531–56.

Leong, K. L. H., M. A. Yoshimura, and H. K. Kaya. 1997. Occurrence of a neogregarine protozoan, *Ophryocystis elektroscirrha* McLaughlin and Myers, in populations of monarch and queen butterflies. Pan-Pacific Entomologist 73:49–51.

Malcolm, S. B. 1993. Conservation of monarch butterfly migration in North American: An endangered phenomenon. Pp. 357–61 in S. B. Malcolm and M. P. Zalucki, eds., Biology and Conservation of the Monarch Butterfly. Natural History Museum of Los Angeles County, Los Angeles.

Malcolm, S. B., B. J. Cockrell, and L. P. Brower. 1989. Cardenolide fingerprint of monarch butterflies reared on common milkweed, *Asclepias syriaca* L. Journal of Chemical Ecology 15:819–53.

———. 1993. Spring recolonization of eastern North America by the monarch butterfly: Successive brood or single sweep migration? Pp. 253–67 in S. B. Malcolm and M. P. Zalucki, eds., Biology and Conservation of the Monarch Butterfly. Natural History Museum of Los Angeles County, Los Angeles.

Marriott, D. F. 1996. Monarchs return to the southland. The Monarch Newsletter 7(2):1.

———. 1999. Overwintering population dynamics of monarch butterflies *(Danaus plexippus)* in northern Baja California, Mexico. Monarch News 9(7):6–9.

Merino-Perez, L., and P. Gerez-Fernandez. 1996. Status of conservation of the monarch butterfly in Mexico. Report for the Commission for Environmental Cooperation for North America, March, pp. 1–32.

Nabhan, G. P. 1987. The Desert Smells Like Rain. North Point Press, San Francisco.

New, T. R., R. M. Pyle, J. A. Thomas, C. D. Thomas, and P. C. Hammond. 1995. Butterfly conservation management. Annual Review of Entomology 40:57–83.

Pyle, R. M. 1981. The Audubon Society Field Guide to the Butterflies of North America. Knopf, New York.

———. 1983a. Monarch butterfly: Threatened phenomenon. Mexican winter roosts. Pp. 463–66 in S. M. Wells, R. M. Pyle, and N. M. Collins, The IUCN Invertebrate Red Data Book. International Union for Conservation of Nature and Natural Resources, Gland, Switzerland.

———. 1983b. Monarch butterfly: Threatened phenomenon. Californian winter roosts. Pp. 467–70 in S. M. Wells, R. M. Pyle, and N. M. Collins, The IUCN Invertebrate Red Data Book. International Union for Conservation of Nature and Natural Resources, Gland, Switzerland.

———. 1983c. Migratory monarchs: An endangered phenomenon. Nature Conservancy News 34(5):20–24.

———. 1997. The historic flight of Monarch #09727. Monarch News 8(3):1, 3–4.

———. 1998. The biogeography of hope: Why transporting butterflies is a bad idea. Monarch News 8(6):6–7.

———. 1999. Chasing Monarchs: Migrating with the Butterflies of Passage. Houghton Mifflin, Boston.

———. 2001. A link between Western and Mexican monarchs. Sunset (January):8.

Roeske, C. N., J. S. Seiber, L. P. Brower, and C. M. Moffitt. 1976. Milkweed cardenolides and their comparative processing by monarch butterflies *(Danaus plexippus)*. Recent Advances in Phytochemistry 10:93–167.

Russell, R. W., F. L. Carpenter, M. A. Hixon, and D. C. Paton. 1994. The impact of variation in stopover habitat quality on migrant rufous hummingbirds. Conservation Biology 8:483–90.

Sakai, W. 1995. The season that was (not). The Monarch Newsletter 5(6):7.

Scott, J. A. 1986. The Butterflies of North America. Stanford University Press, Stanford, Calif.

Snook, L. C. 1993. Conservation of the monarch butterfly reserves in Mexico: Focus on the forest. Pp. 363–75 in S. B. Malcolm and M. P. Zalucki, eds., Biology and Conservation of the Monarch Butterfly. Natural History Museum of Los Angeles County, Los Angeles.

State and Federal Monarch Activities in the United States. 1997. U.S. Department of the Interior, Office of the Deputy Assistant Secretary, Policy and International Affairs, Washington, D.C.

Swengel, A. B. 1994. Population fluctuations of the monarch *(Danaus plexippus)* in the 4th of July butterfly count 1977–1994. American Midland Naturalist 134:205–14.

Taylor, O. 1996. Monarch populations: Fall 1995–spring 1996. Monarch Watch 4(1):4–5.

Urquhart, F. A. 1960. The Monarch Butterfly. University of Toronto Press, Toronto.

Urquhart, F. A., and N. R. Urquhart. 1977. Overwintering areas and migratory routes of the monarch butterfly (*Danaus p. plexippus*, Lepidoptera: Danaidae) in North America, with special reference to the western population. The Canadian Entomologist 109:1583–89.

Vaccaro, R. 1996. Docents flourishing. Friends of the Monarchs 8(12):2.

Wassenaar, L. I., and K. A. Hobson. 1999. Natal origins of migratory monarch butterflies at wintering colonies in Mexico: New isotopic evidence. Proceedings of the National Academy of Sciences USA 95:15436–39.

Wells, S. M., R. M. Pyle, and N. M. Collins. 1983. The IUCN Invertebrate Red Data Book. International Union for Conservation of Nature and Natural Resources, Gland, Switzerland.

Wilbur-Cruce, E. A. 1991. A Beautiful, Cruel Country. University of Arizona Press, Tucson.

Woodson, R. E., Jr. 1954. The North American species of *Asclepias* L. Annals of the Missouri Botanical Garden 41:1–211.

Zahl, P. A. 1963. Mystery of the monarch butterfly. National Geographic 123(4):588–98.

Zedillo, President Ernesto. 2000. Decreto por el que se declara área natural protegida, con el carácter de reserva de la biosfera, la región denominada Mariposa Monarca, ubicada en los municipios de Temascalcingo, San Felipe del Progreso, Donato Guerra y Villa de Allende en el Estado de México, asi coo en los muncipios de Contepec, Senguio, Angangueo, Ocampo, Zitácuaro y Apor en el Estado de Michoacán, con una superficie total de 56,259-05-07.275 hectáreas. Mexico, D.F. (Secretaria de Medio Ambiente, Recursos Naturales y Pesca, SEMARNAP), Nov. 10, 2000, Diario Oficial Gazette, Primera Seccion, pp. 6–29.

CHAPTER 9

The Monarch Butterfly Biosphere Reserve, Michoacán, Mexico

ROBERTO SOLÍS CALDERÓN, IN COLLABORATION WITH BLANCA XIOMARA MORA ALVAREZ, JAIME LOBATO REYES, ELIGIO GARCÍA SERRANO, AND HÉCTOR SILVA RODRÍGUEZ

Conservation of the rich natural heritage of Mexico is commonly a marriage of traditional cultural practices and ecological values that have been handed down through many generations with modern conservation management tools. This blend of traditional and modern knowledge can be seen in the use of forest resources and their associated biodiversity, in the systematic ethno-classifications of aquatic and terrestrial biota, in the development of ecotechnologies that allow diverse agrohabitats, and in the deliberate conservation of energy consumption. Through this marriage of traditional and contemporary approaches to ecological problems, Mexico has managed to maintain its position as one of the four most biologically diverse countries on earth.

The first mandate for environmental protection in Mexico came in the 1880s in the "Desert of the Lions" (near Mexico City)—Mexico's first national park, designated in 1917. This and other early protection measures showed that the Mexican government had made a commitment to conserve and promote landscapes rich in biological diversity. With the creation of the National System of Protected Areas in 1983, quality of habitats became a management priority, with emphasis on both biodiversity and unique natural processes.

The Mexican model of natural heritage conservation has its roots in traditional values and practices, and it is possible to implement only through agreements among willing stakeholders, including rural workers and indigenous peoples, civil and civic organizations, academic research centers, and institutions in the three arms of the Mexican government. An alliance of stakeholders has a greater probability of success in reducing environmental degradation if it also contributes to improvements in the quality of life in local communities.

Given this framework, Mexico has embraced the overwintering region of the monarch butterfly as both a natural and a cultural legacy, a historical gift that has been passed on to the present generation to safeguard. Many of the conflicts that have arisen during the debate over economic sustainability and the environment in Mexico are manifest in this region. The monarch's protected area is in a region of high poverty, and this raises one of the most fundamental issues that touch our society today: how to meet human socioeconomic needs while preserving the environment for the future. The indigenous area in which the monarch overwinters is one in which the rural poor live surrounded by an unparalleled wealth of biodiversity. Its global significance is enhanced by the monarch migration itself, a natural phenomenon shared by Mexico, Canada, and the United States, Mexico's partners in "free trade." Mexican society has many concerns regarding monarchs and the landscape where they overwinter. There are strident and outspoken voices and demands heard from social, economic, and community organizations and from government agencies, academic institutions, community businesses, and other public and individual interests, all of whom marvel at and express continued interest in this migration phenomenon.

In the overwintering region of the monarch, it is urgent that the local population be able to overcome its poverty through the planned and rational use of its own resources. Only through linking human communities with natural communities can we create a relationship that fosters sustainable resource utilization. The monarch's dilemma is not restricted to just the issue of migratory corridors and hibernation sites; it suffers in a broader sense from the impacts of forest and farm management throughout its range. It is necessary to find land-use alternatives such that the people will benefit from this migratory phenomenon rather than be an obstacle to it. In particular, they need to more equitably benefit from the ecotourism that already exists. In addition to tourism, their farming and fishing systems require sustainable techniques and better organization. Forest management that allows ecological restoration in degraded habitats is needed, especially in the oyamel fir habitats. Ecosystem services that result from these conservation efforts will benefit the local human populations.

Monarch Butterfly Overwintering Grounds

The protected monarch overwintering area is in the state of Michoacán, in central-west Mexico, in the physiographic province of the Transvolcanic Axis west of Mexico City (19°19'30" to 20°00'00" latitude and 100°05'30" to 100°20'15" longitude). This fragmented mountain corridor comprises a cluster of hills and volcanic ridges broken in the north and south by plains and valleys of the intermountain region.

The region is characterized by a temperate sub-humid climate with precipitation occurring mostly in the summer. Annual mean temperatures range from 8°C to 22°C, mean precipitation varies from 700 mm to 1,250 mm, and minimum temperatures in the coolest months range from −3°C to 18°C. The climate changes dramatically over relatively short horizontal distances in response to elevation, slope orientation, and abrupt topographic changes that produce distinctive microclimatic conditions and marked local humidity and precipitation gradients. These microclimates are extremely important to the survival of overwintering monarchs.

The flora is transitional between the Nearctic and Neotropical and is considered part of the Southern Mountain Province of Mexico. This montane Meso-American flora displays an enormous biodiversity. The complex combination of abiotic factors and historic biogeographic influences have generated five principal vegetation types:

OYAMEL FOREST (3,600 TO 2,400 M). Key to monarch overwintering, this was formerly the most common vegetation type in the area. It occupies extensive areas of the mountain subsystem and the volcanic ridges. It is characterized by the predominance of the oyamel fir *(Abies religiosa)*, which provides monarchs with their roosts. In disturbed areas, especially on hillsides, the structure of the forest opens and an understory of small trees and shrubs becomes important, including oak *(Quercus)*, alder *(Alnus)*, madrone *(Arbutus)*, willow *(Salix)*, and wild cherry *(Prunus)*, among others. Beneath are subshrub and herbacious layers containing junipers *(Juniperus)*, groundsels *(Senecio)*, *Eupatorium*, *Stevia*, and *Archivacharis*. Mosses such as *Thuidium* and *Minium*, as well as basidiomycete fungi dominate the ground cover during the rainy season.

PINE-OYAMEL FOREST (3,000 TO 2,400 M). This vegetation type has an extensive distribution throughout the region. This mixed forest harbors a diverse flora situated in four distinct layers. Oyamel fir and several pines constitute the top canopy layer. The second stratum is dominated by *Arbutus grandulosa, Salix paradoxa, Alnus firmifolia*, and oaks. The herbaceous layer has diverse elements: *Senecio prenanthoides, S. tolucanus, S. sanguisorbe, Acaena elongata, Oxalis* spp., *Geranium* spp., *Satureja macrostema, Salvia elegans*, and *Asplenium monanthes*, among others. Covering the ground are mosses and wildflowers such as *Viola* spp., *Sibthorpia pichinchensis, Oenothera* spp., and *Oxalis* spp. Among the fungi are *Amanita muscaria, Lactarius sanguifluus, Russula brevipes, Lycoperdon* spp., *Gomphus* spp., *Ramaria* spp., *Clavariadelphus truncatus, Morchella* spp., *Helvella crispa*, and *Boletus* spp. Some of these species are important in the domestic economy as medicinal herbs or foods sold in the local markets.

PINE FOREST (3,000 TO 1,500 M). Most of this forest type occurs in isolated canyon patches that are associated with areas of high relief and relatively constant humidity. In riparian areas within canyons under 2,000 m, the dominant pine is *Pinus pseudostrobus*. In areas with shallow soils and drier conditions, *P. rudis* and *P. teocote* are most common. In addition, *P. oocarpa* and *P. Michoacána* occur in the middle and lower reaches of the zone where canyons are bounded by steep slopes.

OAK FOREST (USUALLY UNDER 2,900 M, BUT TO 3,100 M IN SOME SITES). This forest type is profoundly influenced by succession processes following human disturbances, where mature vegetation rarely persists. Small patches of cypress (*Cupressus* spp.) predominate. The dominant arborescent species are *Quercus laurina, Clethra mexicana, Alnus firmifolia, Salix paradoxa, Buddleia cordata, Buddelia parviflora*, and *Ternstroemia pringlei*. To a lesser extent than in other types, one can find *Abies religiosa, Cupressus lindleyi*, and *Pinus Michoacána*.

CEDAR FOREST (2,400 TO 2,600 M). This forest type is restricted in distribution but shares some landscapes with oak forests. It is represented in the Cerro Pelon Sanctuary, where high humidity and low temperature are the prevalent conditions in riparian zones. *Cupressus lindleyi* and *Abies religiosa* dominate the canopy; *Alnus firmifolia, Salix paradoxa, Senecio angulifolius*, and *Eupatorium* spp. are represented in the next canopy layer. Epiphytes, such as *Usnea barbata*, are abundant on cypress branches, giving this vegetation type a distinctive look and feel.

The Monarch Butterfly Biosphere Reserve has biological importance not only for monarchs but also for its overall species richness, which includes a reported 493 vascular plants; 118 birds, 56 mammals, and 49 fungal species have been reported. The protected area harbors the following endemic species: *Acer negundo* var. *mexicanum, Pinus martinezii, Parus sclateri, Campylorhynchus megalopterus, Pipilo foscus* and *Toxostoma curvirostre*. Federally listed endangered species found there include *Amanita muscaria, Carpinus caroliana, Pseudoeurycea rubertsi, Falco peregrinus*, and *Parabuteo uninctus*. Additional rare species include *Gentiana spathacea, Henicorhina leucophrys, Myoborus miniatus*, and *Dorichs eliza*. Certain other species have special protection status, including *Juniperus monticola, Cantarelus cibarius, Buteo jamaicensis, Denrrortix macroura*, and the monarch itself.

The Migratory Monarch Butterfly

The migratory subspecies of monarch butterfly was the principal reason for declaring this region a Protected Natural Area. Of all the known subspecies of monarchs only one, *Danaus plexippus plexippus*, exhibits the phenomenon of annual migration. The migrations to overwintering grounds in Michoacán and in California are the most notable.

In Mexico, there are two monarch subspecies: *D. p. plexippus* and *D. p. curassaviacae*, the latter being non-migratory. However, both subspecies appear to overlap in central Mexico, including in the overwintering area. Mechanisms for reproductive isolation are in place to avoid hybridization. *D. p. curassaviacae* populations inhabit the high plateau and the mountains of central Mexico during the spring and in the summer, where they live at elevations between 2,000 and 2,700 m. In autumn, this subspecies initiates its descent to elevations below 1,700 m, mostly in the Río Balsas lowlands.

Not long after the descent of *D. p. curassaviacae, D. p. plexippus* populations arrive and form colonies between 3,100 and 2,400 m. In this manner, a temporal and altitudinal allopatric distribution is maintained that impedes the mixing of the two subspecies. In addition, the sexual organs of the migratory monarch are immature at the time of their arrival in the winter refuges, suggesting physiological divergence of the two subspecies.

The migration of the monarchs from southern Canada and the northern United States follows several routes. The destinations for one set of migrants is the Pacific Coast between San Francisco and San Diego in California, and formerly in Baja California as well. The second route utilized by some monarchs begins in southeastern Canada and the northeastern United States and continues through Florida, Cuba, and the Yucatán peninsula. Since overwintering colonies have not been located in the Yucatan, it is assumed that some (or most) continue their journey and fly westward to Michoacán.

The third route originates in the Great Lakes region, in the central United States and Canada. These monarchs enter Mexico near the border towns of Piedras Negras (Coahuila) and Nuevo Laredo (Nuevo León), occasionally as far west as El Paso (Texas) and Agua Prieta (Arizona). Most then move south along the eastern slope of the Sierra Madre Oriental toward Monterrey, Saltillo, and Ciudad Victoria, then on to San Luis Postosí; others follow the Sierra Madre Occidental. Many reach the altiplano (high plateau) by the second half of October. Once there, the route flows to the southeast toward Querétaro. Between the first and third week of November, monarchs begin to arrive at the overwintering sites of Bosencheve, El Oro, Amealco, Atlacomulco, San Felipe del Progreso, Angangueo, and Maravatio in the states of Mexico and Michoacán. Following this route, some monarchs travel up to 5,000 km in about twenty-five days.

As December begins, waves of arriving groups congregate with those that have already arrived, forming fleeting population clusters that sweep up over the mountaintops and through the hollows, continually changing roosting localities. The monarchs begin to seek out roosts that have microclimates suited to their overwintering needs and situate themselves in the high parts of the hollows around 3,200 m. Formerly isolated groups begin to cluster in larger numbers at the best roosts, as roost numbers decrease and roost size increases.

By mid-December, the monarchs have begun their hibernation. They tend to select the most well preserved areas in the oyamel forest, especially on high mountains with southern or southeastern exposures. This offers them the greatest number of hours of exposure to sunlight during the day, allowing them to take advantage of solar radiation on tree trunks. They situate themselves on the middle levels of the trunks of the oyamel firs, where the rush of the low winds and the temperatures are moderated. Here millions of butterflies meet together in seven major and several minor roosts, forming a polygon of mountainous areas decreed as protected in 1986. Here they stay in torpor much of the winter, occasionally descending from trees to seek nectar or water. They are vulnerable to massive die-offs associated with winter snows, like the one suffered in 2001–02 that killed more than half the individuals in some roosts.

As the overwintering season comes to an end in late February, the butterflies migrate down in elevation into the ravines where mixed pine and oyamel fir forests open up. Nectar is more available in these lower elevations at this time where a diversity of plant species begin to flower. The monarchs begin to pair up along the running streams, with ovulation typically beginning in the last week of February. They are capable of producing fertile eggs and mating in March.

With the arrival of the spring equinox, the overwintered monarchs leave their refuge and disperse. They begin their return to the north, laying their eggs as they go. The larvae that result from these broods are nourished primarily by three larval host plant species of the milkweed genus *Asclepias*. The descendents of this generation will be the first to complete their life cycles in northern Mexico, where they will reach their sexual maturity by the end of May or the beginning of June. This first population will either remain in Mexico or pass to more northerly sites where the next several generations will continue their journey northward. Subsequently, their progeny will produce the next three or four generations of monarchs that finally establish themselves in the northernmost regions of the United States and in southeastern Canada, always within the range of the *Asclepias* milkweeds. From the beginning of overwintering until the arrival in the far north of the continent, five to six generations of butterflies will have prospered. Of these, the last of them will emerge primarily in September. By the beginning of October, this last generation of butterflies will follow a route they themselves have never traveled in order to reach the overwintering grounds in Mexico or California.

Challenges to Conservation of Monarch Butterfly Migrations

The principal problems facing monarchs in the region are the typical conflicts that exist between conserving natural habitats (in particular, mature oyamel and oyamel-pine forests) and the urgency to increase the income of human residents in the area. The conservation of natural habitats ensures that monarch overwintering roosts will remain available. However, an important key to maximizing the return of large numbers of migrants and also improving the quality of life of the people that live in the region will be to ensure that the economic benefits of ecotourism can be equitably distributed among the local communities.

In general, five problems can be identified that drive natural habitat deterioration in the area: (1) the human poverty that exists in Michoacán; (2) the strong pressure to consume local natural resources, related to the previous problem; (3) the demise of the local food production system and sales of local products; (4) the poor coordination of different conservation institutions, especially those based on the need for tri-national cooperation; and (5) the lack of participation of local communities in the decision-making process needed to implement conservation measures and sustainable development.

Because of these stresses, the staff of the Protected Area decided to focus their work on containing environmental deterioration, combating poverty, and promoting social and technical advances that achieve sustainable development of the region. To reach these goals, an Interagency Operating Group for the Monarch Butterfly Protected Area was established, for the purpose of coordinating various agencies' local efforts through the Natural Resources and Environmental Secretary (Secretaria de Medio Ambiente, Recursos Naturales y Pesca). This group is working on specific tasks using different democratic regional planning tools and strategies under Mexican environmental law.

The priority actions for the period 1998–2005 focus on environmental conservation and restoration. Projects will take on zoning and regulation, reforestation, soil conservation and aquifer recharge, fire prevention, oyamel fir transplanting and seed production, monarch butterfly monitoring, and biodiversity inventory. At the same time, environmental education workshops will be held to train rangers and increase community awareness. One key effort is training indigenous technicians in natural resources management techniques in the Monarch Butterfly Biosphere Reserve using a manual of basic information on monarch conservation, while setting up an integrated system for information databases and data distribution via the World Wide Web. In addition, interpretive signs for tourists are being installed. Finally, an action plan for sustainable development is being implemented based on community participation through collaboration with the Technical Advisory Council, the Program for the Indigenous Nations, the Regional Sustainable Development Councils, and local watchguard committees.

In short, our approach incorporates the cultural diversity of the inhabitants of the region and allows us to use a variety of nature-society relationship models. These include the traditional ecological knowledge of the Mazahua and Otomi tribes of this region, as well as the management and multiple-use strategies of natural resources by the various rural communities and organizations. Additionally, we are trying to incorporate sustainable urban consolidation processes with new, nontraditional economic activities such as ecotourism. These will be used to enhance the complex social environment and to develop a natural protected-area model of resource management and conservation that responds to both the biological and the cultural diversity that characterizes the region.

APPENDIX: RESUMEN

En la región de la Monarca urge el desarrollo de su población para poder superar la pobreza a partir del uso planificado y racional de sus propios recursos, para que en la vinculación de los grupos sociales con los ecosistemas, se construya una relación de sustentabilidad que haga posible un aprovechamiento renovable. La Monarca y sus problemas no se restringen a la conservación de los corredores y sitios de hibernación, ni a los impactos del manejo forestal dentro de ellos, en realidad se requiere de alternativas regionales para que la gente se beneficie del fenómeno migratorio, el cual no debe ser un obstáculo para su desarrollo, se hace necesario el impulso ordenado del ecoturismo, el establecimiento de sistemas agrícolas intensivos y orgánicos, la piscicultura, el manejo forestal que permita la restitución de los bosques templados y en especial los de Oyamel y de Pino-Oyamel, el diseño de programas productivos y sociales integrales y el financiamiento suficiente y oportuno para que la población local se beneficie de los servicios ambientales que resultan de la conservación.

El área sujeta a protección se ubica en el centro-occidente de la República Mexicana, en la Provincia Fisiográfica del Eje Volcánico Transversal, entre las coordenadas geográficas: 19° 19' 30" y 20° 00' 00" de latitud norte y los 100° 05' 30" y 100° 20' 15" de longitud oeste. Se trata de un sistema montañoso discontinuo, compuesto por un conjunto de serranías y aparatos volcánicos agrupados en la porción centro-norte y separados al norte y sur por planicies y valles intermontanos, las máximas elevaciones corresponden a el Cerro Altamirano, 3 320 msnm. en la parte mas norte, Campanario, 3 640 msnm., Cerro El Mirador 3 340 msnm., Huacal, 3 200 msnm., Chivatí 3 180 msnm. y Cerro Los Madroños 3 040 msnm., en el corredor Sierra Chincua-Campanario-Chivatí-Huacal, en la porción sur destacan, Cerro Pelón, 3 500 msnm., Cacique, 3 300 msnm., El Piloncillo 3,300 msnm. y Cerro La Palma 3 300 msnm.

Desde el punto de vista florístico el área forma parte de una zona de transición entre las regiones Neártica y Neotropical adscrita a la provincia de las Serranías Meridionales, pertenecientes a la Región Mesoamericana de Montaña, lo cual se expresa en la enorme biodiversidad de la zona. La complejidad del conjunto de elementos abióticos y los procesos biogeográficos ha dado lugar a cinco principales tipos de vegetación: el bosque de Oyamel, el bosque de Pino y Oyamel, el bosque de Pino, el bosque de Encino y el bosque de Cedro. El Área Natural Protegida Mariposa Monarca también adquiere singular relevancia si se considera la biodiversidad de especies que posee, se tienen registradas 493 especies de plantas vasculares, 184 especies de vertebrados de los que 118 son aves y 56 mamíferos y 49 especies de hongos, asimismo destaca la presencia en este sitio de especies endémicas, amenazadas, raras y especies bajo protección especial como: *Amanita caesarea, Juniperus monticola, Ambistoma ordinarium* y *Buteo jamaicensis*.

Una de las especies que se encuentran bajo protección especial y que ha sido el factor principal para declarar esta zona como Área Natural Protegida es precisamente la Mariposa Monarca, posee una distribución muy amplia, se le encuentra desde Canadá hasta el sur de Perú, sin embargo, se sabe que existen diferentes subespecies de la cual sólo una es la que presenta el fenómeno migratorio anual y es la que se denomina *Danaus plexippus plexippus*.

Año con año, coincidiendo con el equinoccio de otoño, las monarcas abandonan sus hábitat de verano en la región de los Grandes Lagos, al noreste de Estados Unidos y sureste de Canadá. A mediados de octubre penetran en México por Ciudad Acuña, Piedras Negras y Nuevo Laredo, pasan por la vertiente interna de la Sierra Madre Oriental a la altura de Monterrey, Saltillo y Ciudad Victoria, buscan las montañas bajas de San Luis Potosí, por donde ingresan hacia el altiplano, una vez ahí, la ruta fluye con dirección suroeste acercándose a Querétaro. Entre la primera y tercera semana de noviembre, las monarcas se acercan paulatinamente a los sitios de hibernación desde Bosencheve, El Oro, Amealco, Atlacomulco, San Felipe del Progreso, Angangueo, Maravatío y otras localidades de los estados de México y Michoacán. A lo largo de ésta ruta las monarcas recorren aproximadamente cinco mil Kilómetros, en alrededor de 25 días.

Para mediados de diciembre las monarcas han iniciado la hibernación, para ello seleccionan los sitios del bosque de Oyamel mejor conservados, en laderas de las montañas con exposición al sur o suroeste que les proporcionan mayor número de horas luz durante el día, así como mejores condiciones para aprovechar la incidencia de la radiación solar y se sitúan en las partes medias de los troncos de oyameles en donde el flujo del viento baja y las temperaturas son más benignas. Al termino de la temporada de hibernación, hacia finales de febrero, las mariposas descienden a las cañadas cubiertas por bosques de Pino y Oyamel en donde la disponibilidad potencial de néctar es mayor, tanto porque ha iniciado la floración como por la abundancia de arbustos y se ubican en parajes cercanos a corrientes superficiales de agua, para entonces han madurado sexualmente y realizan el apareamiento. Después de pasar el invierno en México y haber dado inicio a su fase reproductiva, la generación migratoria sale del refugio, coincidiendo con el equinoccio de primavera y se dispersa para su regreso hacia el norte del continente ovopositando generalmente en la zona centro-sur y al este de los Estados Unidos.

Para fines de conservación y protección el área tiene dos decretos que están en vigor y que establece, el primero publicado en 1980, una Zona de Reserva y Refugio de Fauna Silvestre en los lugares donde la Mariposa Monarca hiberna y se reproduce y un segundo decreto de 1986, que establece cinco polígonos con una superficie total de 16 110 hectáreas para fines de migración, hibernación y reproducción de la mariposa monarca, así como la conservación de sus condiciones ambientales. Estos cinco santuarios corresponden a: Cerro Altamirano, Sierra Chincua, Sierra el Campanario, Cerros Chivatí-Huacal y Cerro Pelón.

SELECTED REFERENCES

Alonso, M. 1993. Estudio de la vegetación que comprende el hábitat de invernación de *Danaus plexippus* (Mariposa Monarca) de la "Reserva Especial de la Biosfera Mariposa Monarca." Tesis, Escuela Nacional de Estudios Profesionales Iztacala. Universidad Nacional Autónoma de México.

Alonso, M., and A. Arellano. Mariposa Monarca, su hábitat de hibernación en México. Revista Ciencias, Universidad Nacional Autónoma de México.

Brower, L. P. 1995. Understanding and misunderstanding the migration of the monarch butterfly (Nymphalidae) in North America: 1857–1995. Journal of the Lepidopterists' Society 49:304–85.

Brower, L. P., and R. K. Walton. 1996. Report on the status of the Monarch Butterfly *(Danaus plexippus)* in the United States. Borrador final preparado para el Consejo de Cooperación Medioambiental para el Tratado de Libre Comercio de Norte América.

Calvert, W. H., and L. P. Brower. 1986. The location of monarch butterfly (*Danaus plexippus* L.) overwintering colonies in Mexico in relation to topography and climate. Journal of the Lepidopterists' Society 40:164–87.

Chapela, G., and D. Barkin. 1995. Monarcas y Campesinos. Centro de Ecología y Desarrollo, A.C. Multidiseño Gráfico. 1a. ed. México, D.F.

De la Maza, E. R. 1995. Visión Panorámica del Fenómeno de la Mariposa Monarca y su Posible Conservación. Revista Ciencias. Universidad Nacional Autónoma de México.

Diario Oficial de la Federación. 1980. Decreto que declara zonas de reserva y refugio silvestre, los lugares donde la mariposa inverna y se reproduce. 9 Abril. México, D.F.

———. 1986. Decreto que declara áreas naturales protegidas para fines de migración, invernación y reproducción de la Mariposa Monarca. 9 Octubre. México, D.F.

———. 1994. Norma Oficial Mexicana (NOM-059-ECOL). 16 Mayo. México, D.F.

Espejo, S., L. G. Segura Brunhuber, and C. Ibarra. 1992. La Vegetación de la Zona de Hibernación de la Mariposa Monarca (*Danaus plexippus* L.) en la Sierra Chincua. Tulane Studies in Zoology and Botany, Supplementary Publication No. 1 (Biogeography of Mesoamerica).

Instituto Nacional de Ecología. 1995a. La Reserva Especial de la Biosfera Mariposa Monarca. Problemática y Perspectivas. El Colegio de México, México, D.F.

———. 1995b. Memorias Seminario: Taller de la Región de la Mariposa Monarca con Organizaciones no Gubernamentales, Instituciones Académicas y Ejidos y Comunidades. Angangueo, Estado de Michoacán, México, D.F.

Madrigal, X. 1967. Contribución al conocimiento de bosques de oyamel en el Valle de México. Boletín Técnico No. 18. Instituto de Investigaciones Forestales, México, D.F.

Rzedowski, J. 1988. Vegetación de México. Editorial Limusa, México, D.F.

SEMARNAP. 1997. Estrategia integral para el desarrollo sustentable de la región de la Mariposa Monarca. SEMARNAP, México, D.F.

Contributors

Joaquín Arroyo, Instituto Nacional de Antropología e Historia, Moneda No. 16, Centro Delegación, Cuauhtémoc, D.F. 06060, Mexico.

Lincoln P. Brower, Sweet Briar College, Sweet Briar, VA 24595.

William A. Calder (deceased) University of Arizona, Department of Ecology and Evolutionary Biology.

Sarahy Contreras-Martínez, Instituto Manantalán de Ecología Y Conservacion de la Biodiversidad, Departamento de Ecología y Recursos Naturales, Centro Universitario de la Costa Sur, Universidad de Guadalajara, Autlán de Navarro, Jalisco, Mexico 48900.

Theodore H. Fleming, Department of Biology, University of Miami, Coral Gables, FL 33124.

Eligio García Serrano, Reserva de la Biosfera Mariposa Monarca, Rey Caltzontzin No. 670, Colonia Felix Ireta, Morelia, Michoacán 58070, Mexico.

Russell A. Haughey, Arizona Game and Fish Department, 7200 E. University Drive, Mesa, AZ 85207.

Karen Krebbs, Arizona-Sonora Desert Museum, 2021 N. Kinney Road, Tucson, AZ 85743.

Jaime Lobato Reyes, Reserva de la Biosfera Mariposa Monarca, Rey Caltzontzin No. 670, Colonia Felix Ireta, Morelia, Michoacán 58070, Mexico.

Carlos Martínez del Rio, Department of Zoology and Physiology, University of Wyoming, Laramie, WY 82071-3166.

Rodrigo A. Medellín, Instituto de Ecología, UNAM, Ap. Postal 70-275, 04510 México, D.F.

Blanca Xiomara Mora Alvarez, Reserva de la Biosfera Mariposa Monarca, Rey Caltzontzin No. 670, Colonia Felix Ireta, Morelia, Michoacán 58070, Mexico.

Gary Paul Nabhan, Northern Arizona University, Center for Sustainable Environments, Post Office Box 5756, Flagstaff, AZ 86011.

Robert M. Pyle, 369 Loop Road, Gray's River, WA 98621.

Ana Lilia Reina G., Arizona-Sonora Desert Museum, 2021 N. Kinney Road, Tucson, AZ 85743.

Irma Ruan-Tejeda, Instituto Manantlán de Ecología y Conservación de la Biodiversidad, Departamento de Ecología y Recursos Naturales, Centro Universitario de la Costa Sur, Universidad de Guadalajara, Autlán de Navarro, Jalisco, Mexico 48900.

Ruth O. Russell, University of Arizona, Department of Ecology and Evolutionary Biology, Post Office Box 210088, Tucson, AZ 85721.

Stephen M. Russell, University of Arizona, Department of Ecology and Evolutionary Biology, Post Office Box 210088, Tucson, AZ 85721.

Eduardo Santana C., Instituto Manantlán de Ecología y Conservacion de la Biodiversidad, Departamento de Ecología y Recursos Naturales, Centro Universitario de la Costa Sur, Universidad de Guadalajara, Autlán de Navarro, Jalisco, Mexico 48900.

Jorge E. Schondube, Centro de Investigaconies en Ecosistemas (CIEco), Universidad Nacional Autónoma de Mexico, Antigua Carretera a Patzcuaro No. 8701, Col. Ex-hacienda de San Jose de la Huerta, Morelia, Michoacán 58190, Mexico.

Héctor Silva Rodríguez, Reserva de la Biosfera Mariposa Monarca, Rey Caltzontzin No. 670, Colonia Felix Ireta, Morelia, Michoacán 58070, Mexico.

Roberto Solís Calderón, Reserva de la Biosfera Mariposa Monarca, Rey Caltzontzin No. 670, Colonia Felix Ireta, Morelia, Michoacán 58070, Mexico.

J. Guillermo Téllez, Instituto de Ecología, UNAM, Ap Postal 70-275, México, D.F. 04510.

Thomas R. Van Devender, Arizona-Sonora Desert Museum, 2021 N. Kinney Road, Tucson, AZ 85743.

Blair O. Wolf, University of Arizona, Department of Ecology and Evolutionary Biology, Post Office Box 210088, Tucson, AZ 85721.

Index

About the Editors

Gary Paul Nabhan has done field research on agave conservation and use for a quarter century in the United States and Mexico. He is a recipient of a MacArthur "Genius Award" Fellowship and a Lifetime Achievement Award from the Society for Conservation Biology. Nabhan is author or coauthor of sixteen books, three of which have received national or international awards. He is Director of the Center for Sustainable Environments at Northern Arizona University.

Richard C. Brusca is Executive Program Director and Director of Research at the Arizona-Sonora Desert Museum. He is also an Adjunct Scholar at the University of Arizona and the Centro de Investigación en Alimentación y Desarrollo (CIAD) in Hermosillo, Mexico. He is the author of more than a hundred research publications and seven books, including the largest-selling text on invertebrate zoology (*Invertebrates*, Brusca and Brusca) and the much-used *Common Intertidal Invertebrates of the Sea of Cortez* (University of Arizona Press). He has been the recipient of numerous research grants from the National Science Foundation, NOAA, the National Geographic Society, the National Park Service, the Charles Lindberg Fund, and other agencies and foundations. His areas of research include Sonoran Desert natural history, invertebrate zoology, freshwater and marine ecology, conservation biology, arthropod diversification, phylogenetics, and global biodiversity. He received his Ph.D. from the University of Arizona in 1975, and since that time he has served on panels and boards for many foundations and agencies, including the National Science Board, the National Science Foundation, the Smithsonian Institution, NOAA, the Pew Program in Conservation and the Environment, the Slocum-Lunz Foundation, the Public Broadcasting Service, the IUCN Species Survival Commission, and the U.S. Department of the Interior. He has conducted field expeditions on every continent but has maintained his research programs in the Sonoran Desert and the Sea of Cortez for more than thirty years. He is a Fellow in both the American Association for the Advancement of Science and the Linnean Society of London.

Louella Holter has worked since 1982 as a scientific and technical editor at the Bilby Research Center, which is affiliated with the Center for Sustainable Environments at Northern Arizona University.